S0-ATR-897

Discovering the Expanding Universe

The discovery of the expanding universe is one of the most exciting exploits in astronomy. This book explores its history, from the beginnings of modern cosmology with Einstein in 1917, through Lemaître's discovery of the expanding universe in 1927 and his suggestion of a Big Bang origin, to Hubble's contribution of 1929 and the subsequent years when Hubble and Humason provided the essential observations for further developing modern cosmology, and finally to Einstein's conversion to the expanding universe in 1931. As a prelude, the book traces the evolution of some of the notions of modern cosmology from the late Middle Ages up to the final acceptance of the concept of galaxies in 1925.

Written in non-technical language, with a mathematical appendix, the book will appeal to scientists, students and all those interested in the history of astronomy and cosmology.

HARRY NUSSBAUMER is Professor Emeritus at the Institute of Astronomy, ETH Zurich.

LYDIA BIERI is Assistant Professor in the Department of Mathematics at Harvard University.

Discovering the Expanding Universe

HARRY NUSSBAUMER
Institute of Astronomy, ETH Zurich (Switzerland)

LYDIA BIERI
Department of Mathematics, Harvard University (USA)

CAMBRIDGE
UNIVERSITY PRESS

CAMBRIDGE UNIVERSITY PRESS
Cambridge, New York, Melbourne, Madrid, Cape Town, Singapore, São Paulo, Delhi

Cambridge University Press
The Edinburgh Building, Cambridge CB2 8RU, UK

Published in the United States of America by Cambridge University Press, New York

www.cambridge.org
Information on this title: www.cambridge.org/9780521514842

First published 2009

Printed in the United Kingdom at the University Press, Cambridge

A catalogue record for this publication is available from the British Library

ISBN 978-0-521-51484-2 hardback

Contents

Acknowledgements

Several astronomers read early drafts. The comments of Kevin Briggs, Franco Joos and Hans Martin Schmid helped to define the tracks better, which we then followed. We also thank Simon Lilly for pertinent suggestions. We are particularly grateful for extended comments and suggestions of Hilmar Duerbeck, Allan Sandage and Norbert Straumann. We highly appreciate the foreword by Dr Allan Sandage, adding authentic breadth with his very personal impressions of events and personalities, and on Hubble's reluctance to accept the concept of an expanding universe.

We endeavoured to consult the original sources as much as possible, so we often depended on the help and goodwill of librarians. We thank the library of ETH Zürich, the Zentralbibliothek Zürich, the Leiden Observatory Archives, Barbara Wolff of the Albert Einstein Archives of Jerusalem, Liliane Moens of the Archives Lemaître, Jennifer Goldman and Shelley (Charlotte) Erwin of The Huntington Library. We also thank Dr H. C. Carron, as well as the Master and Fellows of Emmanuel College, Cambridge, for permission to reproduce two illustrations from *A prognostication everlasting* by Thomas Digges, and Jim Kaler for providing the spectrum of the planetary nebula NGC 6818.

Foreword

In 1929, Edwin Hubble published a paper that correlated redshifts of galaxies with distances he had estimated from his calibration of their absolute magnitudes previously made in 1926. Writers of both popular accounts and technical textbooks have often described this as the discovery of the expanding universe. It is not so. This meticulously researched book on the history of the discovery traces the complete story of that discovery. The history started a decade before Hubble's, brilliant to be sure, initial correlation, even as it was followed by the further major advance two years later by Hubble and Humason using new observational data (Hubble and Humason 1931). These had, in fact, led to the convincing conclusion that there is indeed a relation between redshift and distance, but not to the reason for a redshift–distance effect, nor to an expanding universe per se. This definitive book, so thoroughly researched for the wider history that started more than a decade earlier than 1929, uses many heretofore-unused original sources, and many others not often cited in other histories.

The misconception is broadly held in the popular press that Hubble discovered the expansion, primarily from the redshift observations of Slipher, which in 1929 Hubble had correlated with distances. Such accounts, if left bare, neglect the central theoretical underpinnings based on Einstein's General Relativity. De Sitter first introduced the notion of cosmological motion in 1917 with his discovery, now considered a mathematical curiosity, of a time-dependent component of a metric satisfying the Einstein equations that contains a factor with distance. The de Sitter prediction was followed in 1922 by Friedmann's more realistic time-dependent metric that is the basis of the modern standard model.

Nevertheless, despite its strangeness, the 'de Sitter effect' spurred many observation astronomers and others to search for the effect among disparate data, for example, by using kinematic data of galactic globular clusters before the discovery of the wider universe of galaxies in the mid 1920s. A proper accounting of the complete history, and in particular of the crucial role of Lemaître in setting out the basis of the theory as we now know it, is the central theme of this important book.

The work stands so solidly on its merits that a foreword by one who is not an author is not needed, and on many grounds would be inappropriate. Why then this foreword? The short answer is that, by the circumstances of history, I became involved with the problem in 1950 as the observing assistant to Hubble at Palomar. I came to understand his purely observational approach to the problem, devoid of much of a theory, and in fact to see his reluctance to believe that the expansion is real. This, of course, is one of the ironies of the history. Because I could describe the Pasadena mood as I learned it, first by working with Hubble and later in re-determining his correction factors to apparent magnitudes to understand the basis of his reluctance, I accepted the opportunity offered by the authors to write this foreword.

Every subject of inquiry can be described and studied on different levels, whether in science, opinion, philosophy, the arts, or all other inventions of the human mind. Each level becomes part of a hierarchy ordered by complexity. It is written, 'It is an essential feature of science that one can analyze a particular level [of the hierarchy] without knowing anything of the lower and the higher levels. It will generally be true that an understanding of the lower and higher levels will enhance an understanding of a particular level, but for some purposes it is not necessary for an understanding of the system at that level.' (Murphy and Ellis 1996.)

The hierarchical approach to the history of cosmology is especially useful here as the subject developed in the middle years of the last century. Initially I knew one level of the hierarchy from the observational side working as Hubble's assistant from 1949 to 1953 in Pasadena. The theoretical levels of the hierarchy had been largely developed elsewhere, and were not needed at that time by the Mount Wilson observers for their level of the work.

As this book describes, the theoretical levels of the modern hierarchy had begun with General Relativity in 1917, twelve years before Hubble's keystone paper of 1929. Predictions amenable to observations followed from de Sitter's discovery in 1917 of his mysterious 'distance dependent time' in the equation of the metric of the differential geometry of spacetime. Astronomers who knew of the predictions, such as Lundmark and Silberstein in 1924 and 1925, had begun to search for the de Sitter effect in the kinematics of many types of astronomical objects before galaxies were understood to define the larger scale structure of the Universe.

The search centred on the problem from classical kinematic astronomy of the 'solar motion' relative to the white nebulae. This search had been started in 1916 and continued to 1925, even before the white nebulae had been proved to be the island universes of the galaxies. Leaders in this were O. H. Truman in the United States, C. Wirtz in Europe (often called the European Hubble without a telescope), and Vesto Slipher in Arizona, who had measured most of the red-shifts that were used in the solar motion solutions.

Even as Slipher was after the solar motion, astute theoreticians such as Eddington had made the connection with the de Sitter effect, throwing the solar motion problem into the realm of cosmology. In 1923, Eddington (1923) had already reprinted Slipher's data in his theoretical book on relativity.

Hubble also was after the solar motion, as his 1929 announcement paper makes clear. Yet at the end of the paper he does mention the de Sitter effect, but not the much more important theoretical underpinning papers by Friedmann in 1922, Lemaître in 1927 and Robertson in 1928, which he could have known in 1929. These, like the mysterious de Sitter metric, were also based on General Relativity. Hubble's 1929 announcement was only to point out that a relation exists between redshifts and distance, but not to claim an understanding of the effect in terms of an expansion. This book provides a definitive discussion on these points.

Furthermore, from Hubble's written records throughout the 1930s, it is clear that the later advances from Pasadena did not depend on the theory developed by Lemaître and Robertson, and subsequently advanced by McVittie, Heckman, Milne and McCrea, and others. These culminated in the Mattig revolution where the relevant equations for the observational approach were put in closed form valid for all redshifts, not in series expansions valid only for small redshift values. Only the most elementary theory was needed to search for the second-order term in the observations, now called the deceleration parameter, q_0 (see Humason et al. 1956 and Roberston 1955).

Following the Mount Wilson developments in the 1930s, the cosmological program for the Palomar 200-inch had been largely set by 1936. At Palomar, it was to be more of the same that had been done at Mount Wilson. Central was to be the improved formulation of the redshift–distance relation extending the redshift range beyond the capabilities of the Mount Wilson telescopes, and a re-calibration of Hubble's 1930s distance scale using magnitude scales set up photoelectrically rather than by photographic methods. The work could again proceed largely without appeal to theory. To be sure, Tolman and later Robertson were in Pasadena at the California Institute of Technology, but they were studying the problem at distinctly different levels of the hierarchy.

Hubble, often interviewed by the press in the late 1930s and again near the completion of the Palomar reflector, invariably replied to questions on the meaning of the redshift that he was not interested in theoretical hypotheticals, but rather on the data to be obtained at high redshifts to test the reality of a true expansion, which he had often argued against.

I began working as Hubble's assistant while still a graduate student in astronomy at CalTech. At the same time, I was observing with Walter Baade at the Mount Wilson 60- and 100-inch reflectors for a thesis problem on globular clusters related to stellar evolution. My work with Hubble started in the summer of 1949 on a program of galaxy counts, to follow up on his early 1934 count program to find the curvature of space based on Eddington's 1932 proposal to measure how spatial volumes increase with distance (Eddington 1933, Chapter 2).

However, Hubble suffered a major heart attack in late summer of 1949. He could not continue his Palomar observational cosmology project at the telescope. The program was considered central enough for the overall Palomar program that the cosmological segment of the work must be continued despite Hubble's temporary (it would prove to be permanent) incapacity to observe. I was given the observing responsibility at Palomar for the Cepheid program and for the photometric program to obtain the apparent magnitudes of the cluster galaxies in Humason's cluster redshift program for the redshift–distance effect.

I had almost daily contact with Hubble from 1951, during which I learned much of his methods at the observational level of the hierarchy. With his death in September 1953, that part of the Palomar cosmological program concerning the distance scale, the extension of the redshift–distance program into the realm of the second parameter, and the galaxy classification program fell to me. With that event, I became a Pasadena observational cosmologist.

But as the work progressed into the 1960s, it became evident that the redshifts measured by Humason were large enough, such that theory had to be brought into discussion if the expansion were real. I also came to understand why Hubble, mistakenly, had doubts on the reality of the expansion, and why, to appreciate his reluctance, it was necessary to employ the full machinery of the Mattig revolution (Mattig 1958, 1959), with which we began to interpret the new observations and perform the tests in the early 1960s (see Sandage 1961, Sandage and Perelmuter 1991, Lubin and Sandage 2001).

Hubble's scepticism on the reality of the expansion was based on his analysis of his observations of both the redshift–distance relation, and his 1934 galaxy counts. From 1936, he had argued that if he applied his corrections for the effects of the redshift on the measured apparent magnitudes in his two observational programs (the Hubble diagram of redshift vs. apparent magnitude, and the galaxy number count–magnitude correlation) and the expansion was real, he would reach a contradiction in both programs. In contrast, if he applied only a correction that denied the expansion, the contradiction would disappear. The details are more complicated than need be set out here, but are explicit in Hubble's 1936 Oxford Rhodes Lectures (Hubble 1937). Until his death in 1953 he kept the door open as to the expansion not being real. This is seen directly in his 1953 Darwin Lecture (Hubble 1953), given only a few months before his death. In the Hubble diagram shown in that published account, there explicitly is no correction for a real recession.

In an auxiliary program, using both the Mount Wilson and the Palomar telescopes, we discovered in the 1960s that his correction terms to magnitudes for the effects of redshift were faulty, based as they were on too hot an effective spectral energy distribution used to calculate the corrections (Oke and Sandage 1968). The same had been found earlier, based on older photographic photometry (Greenstein 1938), but had been ignored at the time. Use of the correct correction terms, plus Robertson's (1938) connective equation between luminosity, proper distance and redshift, eliminated the discrepancy.

The irony, of course, is that although the discovery of the expansion is often attributed to Hubble with his 1929 paper, he never believed in its reality. It was left to the advances after the mid 1950s to establish its reality on many fronts: the Tolman surface brightness test, the agreement of the timescales of stellar evolution with the expansion age using the revised extragalactic distance scale, the time dilation in supernovae light curves with increasing redshifts, the increase of the cosmic background radiation temperature with look-back time at increasing redshift, and the existence of the background radiation itself, redshifted so that the Planck curve has precisely the correct intensity normalisation for its temperature (i.e. its zero chemical potential), satisfying again the Tolman $(1+z)^4$ surface brightness test.

Fate has permitted a career for me in cosmology by the accident of timing. I was of such an age to fit into the transition period between the pioneer Mount Wilson observers of Hubble, Humason, Minkowski and Baade in the two decades from 1930 to 1950, and the modern theoretical cosmologists, plus the new generation of observers using telescopes in space to study what they believe are origins, in earlier times called cosmogony.

During that 50-year period from 1950 to 2000, I became acquainted with many of the players in both the observational and theoretical realms of the old school. My association and strong friendship with Milton Humason lasted well beyond his retirement from the Mount Wilson and Palomar Observatory in 1957 until his death in 1972. H. P. Robertson was my professor of theoretical physics at Caltech in 1950/51. He vetted my 1961 paper (Sandage 1961) on the program for the 200-inch Palomar telescope. It was then that he told me of his conversations with Hubble in 1928 about his theoretical paper on an expanding universe and an evaluation of the expansion rate using Hubble's distance scale of 1926. Fred Hoyle became a colleague at the Mount Wilson and Palomar Observatory in the 1950s and we wrote a paper together on the observational value of the deceleration parameter (Hoyle and Sandage 1956). We also often argued on the validity or otherwise on the Steady State model, which I had written against from the observational side. We remained friends throughout. I knew McVittie well and often discussed with him the observations being made at Palomar on the redshift–distance relation and the determination of the second-order term. His book, *General Relativity and Cosmology*, went through two editions before and after the Mattig revolution. In the first, in 1956, the equations relating redshift to distance and the spatial volumes enclosed therein were all in series expansions of the redshift. The second edition in 1965 set out all the equations in closed form valid for all values of the redshift, no matter how large, based on the 1958/59 Mattig equations.

At the time of the 1967 Prague IAU meeting I had not yet met Mattig. But there I thought I had seen, at a distance, a name badge marked 'Mattig'. My awe of his two revolutionary papers of 1958/59 was too great for me to talk directly with him, so I did not follow up my glance at the badge to make an acquaintance. I learned only much later that Mattig was not even at the Prague meeting, so it has turned out that I was in awe of a mistakenly read badge rather than the real thing. However, in the 1990s we did spend a day together in Basel with G. A. Tammann, my long-time colleague, on the distance scale problem beginning in 1963. We talked about classical cosmology which, at that time, was still centred about 'the search for two numbers' (Sandage 1970), the second of which was based on Mattig's equations. In connection with his Ph.D. examination (his Ph.D. was on sunspots), Mattig had been given the task of presenting a cosmological problem still open at the time. He succeeded in giving a closed analytical solution for the relation between redshift and bolometric luminosity. Out of that exercise came the two key papers of 1958 and 1959 that changed the course of theoretical discussions as far as they relate to observations.

I became acquainted with Heckmann in 1957 at the Vatican conference on stellar populations. We walked throughout Rome during breaks in the conference and had discussions about the history of many of the sites we visited that he knew so well. I met him again in 1961 at the Santa Barbara IAU symposium 15 on cosmology, the conference proceedings of which were edited by McVittie. Some months later I received a gift from him of his well-worn personal copy of his important book, *Theorien der Kosmologie*, published in 1942 in Berlin during the Second World War. The book was then, and is still today, most difficult to find.

I had also met Lemaître at the Vatican conference on stellar populations in 1957, where stellar evolution was only then drawing near to cosmology via the evolutionary corrections to galaxy luminosities at high look-back times. I met him again at the 1961 Santa Barbara cosmology conference. During a noon lunch break, Lemaître reintroduced himself and recalled our conversations at the Vatican conference. After some discussions about progress in the Palomar program we began a wider discussion on the beauty of the Einstein equations and the mystery of cosmology itself. Toward the end, he asked, 'Sandage, can you really envisage curved space and the beauties of Riemannian geometry, so necessary for relativity?' I replied, 'No, Father, I have tried and tried, using all the tricks known to visualise curved space, but my visualisations have so far failed.' Lemaître then sighed and said, 'I understand, but it is a pity because the visualisation is so beautiful. Perhaps it might be best for you to change fields.' He said it gently, like a father to a son.

Beginning in 1917, the road to the discovery of the expanding universe was traveled by many scientists on each side of the hierarchy between theory and observation. This remarkable book gives credit in a fair and neutral way to many who made the journey. It deserves to be studied by future historians because it is authoritative and definitive.

Allan Sandage
Observatories of the Carnegie Institution
California

1

Introduction

The discovery of the expanding universe in the first half of the twentieth century is one of the most exciting exploits in the history of astronomy; it is the main theme of this book.

An intense and occasionally even fierce controversy divided the astronomical world at the beginning of the twentieth century: does the Milky Way represent the whole Universe, or is our Cosmos composed of an enormous or even infinite number of island universes, each of them similar to our own Milky Way? The question was definitely settled in 1925 in favour of island universes, a concept that had been proposed by Immanuel Kant in 1755. Yet, even before the debate was closed, that grand silent Universe had become the playground of theoreticians, who tried to fathom its structure with their abstract mathematics.

Modern cosmology began in 1917 when Albert Einstein published his 'Kosmologische Betrachtungen zur allgemeinen Relativitätstheorie' (Cosmological considerations on the general theory of relativity). In his theory, space and time form a unity: *spacetime*. They are no longer absolute and independent concepts, the structure of spacetime is intrinsically related to the material and energetic content of the Universe: spacetime is structured by gravitation.

More than two thousand years of astronomical observations showed the Universe to be stable and practically immutable in space as well as in time. Thus, when Einstein merged relativity with cosmology, he quite naturally wanted to accommodate a static universe that did not change with time. Yet, gravitation makes matter condense. With the aim of counteracting that force and maintaining the stability of the Universe, Einstein introduced his famous cosmological term, now generally called the cosmological constant. The resulting solution provided bold answers to age-old questions: the Universe is finite, we know its size, and we know how much matter it contains.

However, the idea of a static universe was soon challenged. In 1922, the Russian Alexander Friedmann showed that the most natural solution of Einstein's fundamental equations of General Relativity is a dynamic universe: it may expand or contract.

Redshifts play an important role in cosmology. A prism splits white light into a continuous range of colours from red to blue, as seen in the rainbow. A range of wavelengths defines the portion of light corresponding to a given colour: short wavelengths for blue, longer ones for red. Stars and galaxies consist of chemical elements like hydrogen, oxygen, carbon and iron. When these elements radiate or absorb light they produce in their spectra characteristic patterns of lines at fixed wavelengths. Often the spectra of stars show that the whole set of these patterns is shifted away from the wavelengths expected from our laboratories. The best known explanation for such wavelength shifts is the Doppler shift: if a radiating object is moving towards us, its spectrum is shifted to the blue; if it is moving away from us it is shifted to the red. Since 1912, Vesto Slipher had observed that spectra of spiral nebulae showed very large redshifts.

The first scientific text – backed by theoretical and observational evidence – that explicitly advocated an expanding universe in the sense we see it today was published by Lemaître in June 1927. He did not know about Friedmann. When he theoretically re-discovered the dynamical universe, he combined that theory with observations, and showed that we live in an expanding universe. His principal observational proof came from spiral nebulae, for which he combined Slipher's redshifts with distances, published by Hubble in 1926.

In the autumn of 1927, Lemaître showed his discovery to Einstein. But Einstein would not hear about a dynamically evolving world. In 1922, he had already shrugged off Friedmann's discovery, and he told Lemaître that his model of the expanding universe seemed mathematically correct, but it was physically abominable. Only in 1931, after Eddington and de Sitter had enthusiastically welcomed Lemaître's publication of 1927 as a true breakthrough for cosmology, and Hubble together with Humason had in 1929 observationally confirmed Lemaître's velocity–distance relationship and its implication that galaxies move away from us, did Einstein finally accept the new concept.

Scientific discoveries are never the sole merit of one individual. They have to be seen in a historical perspective. Yet, all the milestones are tied to outstanding individuals. This is no contradiction. Masterstrokes of individuals become possible when a favourable territory has been prepared and a propitious environment provides indispensable support. This was as valid for Newton as for Einstein. The story we have to tell shows that even geniuses can go astray or fail to recognise the obvious solution to a burning problem.

After an introductory, very condensed summary of Ptolemy's *Almagest*, we start our journey through history in the Middle Ages with the revival of astronomy in the West. We concentrate on developments that are relevant for the eventual discovery of the expanding universe. The cultural revival owed much to close contact with Islamic culture, which had preserved and further developed the classical Greek scientific tradition. These contacts were intensified after the re-conquest of Toledo by the Spaniards in 1085. The translation of classical Greek and Islamic authors from Arabic into Latin in the twelfth and thirteenth centuries gave an enormous boost to Western culture.

We begin our history with Johannes de Sacrobosco. He belonged to a group of scholars who took advantage of the Islamic heritage to create a European astronomy. From him we learn how the scientific community of the thirteenth century saw the Universe. Then, in Cusanus, we meet an eminent philosopher of the early Renaissance, whose ideas about the Heavens fundamentally diverged from traditional scholastic thinking. The following generations – including Nicolaus Copernicus, Giordano Bruno, Johannes Kepler and Galileo Galilei – belong to the 'heroic' set of the sixteenth and early seventeenth century who fought to replace the geocentric by the heliocentric view of the world. In that period we also find the first Western description of a galaxy.

Essential ingredients of modern astronomy entered in 1644 with Descartes' concept of a universe in perpetual evolution, and Newton's formulation of the law of gravitation in 1687. Nebulae became a subject of speculation and investigation with Halley's report to the Royal Society in 1716; Kant first proposed the concept of galaxies in 1755. A new era of observational cosmology began with the giant telescopes of Herschel and Rosse, and the spectroscopy of Huggins. After a few decades of calm, photography greatly helped to revive interest in nebulae, and the debate as to whether some nebulae represented island universes was resumed. In 1925, Kant's hypothesis had definitely won the day. By this time, modern cosmology, which had started with publications by Einstein and de Sitter in 1917, was already on its way.

Our principal aim is to bring together the crucial stepping-stones, through which the concept of an expanding universe and the first suggestions of the Big Bang and vacuum energy merged into a coherent picture in the period from 1917 to approximately 1934. Although Slipher's redshifts were an early intriguing observational stimulus, the cosmological debate was largely theoretical initially. The most important participants were de Sitter, Eddington, Einstein, Friedmann, Lanczos, Lemaître, Robertson, Tolman and Weyl. Observational contributors who were directly involved were Hubble, Humason, Lundmark, Slipher, Strömberg and Wirtz.

On our journey we try to cover a large field in a very limited number of pages. For nearly every person, fact or idea mentioned, there exist detailed investigations, scattered in books or specialised publications. To mention them all, and the shades they add, would result in an encyclopaedic volume, which is not our aim. Those interested in historical details will profit from scanning through the volumes of the *Journal for the History of Astronomy*. We give, however, full references to all the sources that helped us to reconstruct the discovery of the expanding universe. Most of them are the original publications that constitute the scientific backbone of the evolving cosmology of the early twentieth century. A mathematical appendix is given for those who would like to have the rudiments of the theoretical arguments immediately available; it cannot, of course, replace a proper introduction to General Relativity.

When citing sources written in German or French we have translated them directly into English if the original is generally easily accessible. Otherwise we give the original text followed by our translation.

2

Cosmological concepts at the end of the Middle Ages

Greek antiquity turned astronomy into a mathematical science. Ptolemy of Alexandria (*c*. AD 100–170) expounded its essence in the *Almagest* (see Toomer 1998). In that work, the philosophical roots (Plato, Aristotle) were merged with the observational and mathematical foundations (Eudoxus, Callipus, Aristarch, Hipparchos). It presents the Universe as a spherical system built around the motionless Earth. The philosophers demanded the motions of the heavenly bodies to be circular and of constant speed. If they appeared to be different, they had to be traced back to the prescribed ideal motions. The planets, which comprised the Moon as well as the Sun, moved in spherical shells, placed concentrically around the Earth. Ordered according to increasing distance the shells contained: the Moon, Mercury, Venus, the Sun, Mars, Jupiter and Saturn. Immediately adjacent to Saturn followed the shell of the stars. It was surrounded by the invisible *primum mobile*, the shell that provided the impetus to keep the system going.

The *Almagest* contains the theory of that structure, as well as detailed instructions on how to calculate the apparent motions of the planets as seen against the background of the stars, 1022 of which were catalogued. With the decay of the Roman Empire, Greek scientific culture, including the *Almagest*, was lost in the western part of Christendom. It was, however, absorbed into Islamic scientific culture.

After the long dormant phase of the Early Middle Ages, when in the Western Christian world practically all classical astronomical knowledge had been forgotten, contact with Islamic culture, particularly after the re-conquest of Toledo in 1085, gave a new impetus to European science. The translation of the *Almagest* from Arabic into Latin provided the techniques to calculate the paths of the planets. The basic cosmological concept, however, was much influenced by St Thomas Aquinas (1227–1274). He worked on a synthesis of Aristotelian philosophy and Christian teaching. There was, of course, a permanent discussion about how much biblical teaching had to be taken literally. A fundamental difference existed between the Aristotelian eternities of the astral world as opposed to the biblical creation of all things within one week. Nevertheless, a semi-official consensus on astronomical truth emerged.

Philosophy in the second half of the sixteenth and the first half of the seventeenth century had to come to terms with Copernicus and his shift from a geocentric to a heliocentric system. But hardly had that been accepted, when in 1644 Descartes widened the finite spherical cosmos into a universe with no bounds and no centre. At least as important, he changed the biblically created, forever immutable heavens into a dynamical universe, where the properties of matter, and not the unpredictable whim of God, were the creative forces that formed stars and planets. One generation later, Newton's *Principia* (first published 1687) gave scientists the tools to explain and predict the paths of planets and the forces that acted between the stars.

Because of its ability to measure time and geographical positions with great accuracy, astronomy in the seventeenth century attracted considerable financial support from kings and governments. It also captured the curiosity of the general public. In the second half of the seventeenth century, if we believe Molière, astronomy became a very popular pastime of Parisian upper-class ladies. *Les femmes savantes* was first performed in 1672, one year before distances in the Solar System were determined to within about 10% of their actual values by French and English astronomers. In that comedy the landlord complains that his wife, his sister, and his daughter had emptied the top floor of his house in order to install an observatory to find out what was happening on the Moon, instead of keeping the house in order and preparing him a decent meal. If the three ladies were seasoned observers, they might even have spotted Andromeda; the first Western report about that galaxy had been published by Marius in 1614. But nebulae had to wait for the eighteenth century to become objects of serious astronomical research.

The spherically closed universe of antiquity and the Middle Ages

The astronomical model of the late Middle Ages was basically the antique geocentric system. The *Tractatus de Sphaera* by John Holywood (1195?–1256) was probably the most important astronomical textbook, and became part of the standard medieval university curriculum. It was copied, printed and reprinted up to and beyond the times of Copernicus. The author was better known by his latinised identity of Johannes de Sacrobosco. *De Sphaera* was written around 1230. Although it was based on the astronomy of Ptolemy and his Arabic commentators, it gave a rather elementary account of spherical astronomy, with none of the sophistications of Ptolemy, such as a proper treatment of epicycles. Epicycles were frequently included as additions by later commentators. The last important edition was by the Jesuit Christoph Clavius, and was first published in Rome in 1570. The large number of further editions, even in the seventeenth century, bears witness to the importance of this textbook of Ptolemaic astronomy long after the publication of Copernicus' *De Revolutionibus* in 1543.

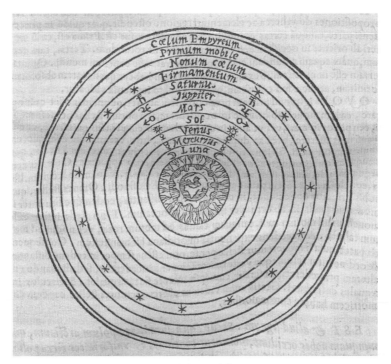

Fig. 2.1 Astronomical cosmos of the Middle Ages. The Earth is in the centre. It is surrounded by water, above water – air, above air – fire. Further still are the guiding spheres of the Moon, the Sun, the planets and the stars. Depending on the author or commentator, there may be several spheres beyond the fixed stars: they take care of additional periodic motions, such as precession, deemed important by the respective author. Often there is just one additional sphere: the *primum mobile*, the unmoved mover, the cause that sets the Universe in motion. In this image, we find beyond the fixed stars (here called *Firmamentum*), the Ninth heaven, the *Primum Mobile* and the *Coelum Empyreum*, the abode of God. (*Sacrobosco in Sphaeram.* Christoph Clavius, Rome, 1581. ETH-Bibliothek, Zurich, Sammlung Alte Drucke.)

The *Almagest* explained the motions of the heavenly bodies, and gave instructions how to calculate them. The most obvious motion is the daily rotation of the sky around the axis through the north and south poles; the celestial equator divides the northern and southern hemispheres. The celestial equator is really an extension of the Earth's equator. We know this apparent motion is due to the daily rotation of the Earth. Because of our annual path around the Sun we see the Sun moving annually on a great circle across the sky. This circle is called the ecliptic. Because of the tilt of the Earth's rotational axis against the plane of the ecliptic, the ecliptic is tilted against the celestial equator by 23.4°. The ecliptic cuts the celestial equator at the two equinoxes, where the Sun crosses the celestial equator in the spring and autumn. The planets move in planes that have only small inclinations against the Earth's orbital plane, with a maximum of 7° for Mercury. The projections of the planetary orbits therefore lie in a small band centred on the ecliptic: the Zodiac.

Observation shows that we are dealing with periodic motions. The daily period is due to the Earth's rotation, the annual period originates in the annual orbit of the Earth around the Sun, and there are the periods due to the planetary orbits around the Sun. There is an additional long-term period of 25 780 years due to the precession of the Earth's axis. Since Newton, we have known that the gravitational forces of the Sun and the Moon acting on the Earth's equatorial bulge cause the precession. Hipparchos had discovered this slow precessional shift of the equinoxes along the ecliptic of approximately 1.4° per hundred years in approximately 130 BC.

For us it is fairly easy to visualise these motions and their effects on the apparent paths of the celestial bodies. However, observers in antiquity and the Middle Ages had to explain apparent motions within the model of an immobile Earth in the centre of a rotating universe. The *Almagest* gave this explanation and the book of Sacrobosco repeated it, but on a very much simplified level. *De Sphaera* was not simply copied. Many different editors augmented it with comments. In this way each generation enriched the cultural heritage; but we have here neither the time nor the space, nor the detailed knowledge to do justice to all the individuals who contributed to it. Instead, we can summarise a few of Sacrobosco's essential points; our source is Thorndike's translation of 1949.

Sacrobosco states that the Universe is divided into two, the ethereal and the elementary regions, i.e. heaven and Earth. The elementary region, existing subject to continual alteration, is divided into four. The Earth is placed at the centre in the middle of all, above which is water, above water is air, above air fire, which is pure and not torpid and reaches the sphere of the Moon. Revolving around the elementary region with continuous circular motion is the ethereal, which is lucid and immune from all variation in its immutable fifth essence. The ethereal is divided into the following: the ninth sphere, which is called the 'first moved' or the *primum mobile*; the sphere of the fixed stars, which is named the 'firmament'; and the seven spheres of the seven planets: the Moon, Mercury, Venus, the Sun, Mars, Jupiter and Saturn. Each of these encloses its inferior spherically.

Two movements have to be distinguished. One is of the outermost heaven. Then there is another movement, oblique to this and in the opposite direction, of the inferior spheres on their axes, distant from the former by 23 degrees. But the first movement carries all the others with it in its rush about the Earth once within a day and night, although they strive against it, as in the case of the eighth sphere, one degree in a hundred years. This second movement is divided through the middle by the Zodiac, under which each of the seven planets has its own sphere, in which it is borne by its own motion, contrary to the movement of the sky, and completes it in varying lengths of time – in the case of Saturn in 30 years, Jupiter in 12 years, Mars in 2 years, the Sun in 365 days and 6 hours, Venus and Mercury in about the same time and the Moon in 27 days and 8 hours.

In chapter two, Sacrobosco comes back to the two movements. 'Be it understood that the first movement means the movement of the primum mobile, that is, of the ninth sphere or last heaven' and 'The second movement is of the firmament and planets'. Thus, the ninth sphere not only provided the fundamental motion for the lower spheres but was also the background for the slow precession of the eighth sphere.

Some astronomers thought precession to be variable (the variability was called trepidation). But, according to Platonic–Aristotelian dogma, all observed motions had to be composed of regular motions, so it was thought necessary to introduce an additional sphere for trepidation. Depending on whether trepidation was explicitly taken into account, the number of shells varied. Thus, in some editions of Sacrobosco we find nine spheres, whereas in others we might find ten or eleven. The eleventh would then be a purely theological sphere: the Empyreum, the immovable place for God and his elect – cosmology in those days was no less complicated than it is today.

There are two fundamental points we should keep in mind about the cosmological concept of the Late Middle Ages: (1) The Earth is a sphere and is located in the middle of the firmament, where it remains immobile. Contrary to what one still frequently reads, people of the Middle Ages knew very well that the Earth has a spherical shape, and that was never a serious theological problem. (2) There is an important division between the earthly sphere in continual alteration and the heavenly immutable confines. It corresponded to the religious imagery of those times, as embodied in Dante's *Divine Comedy*: the corrupt Earth with Hell at its centre and its ever-consuming fire for the damned, whereas the saved ones are with God in the highest sphere where everlasting perfection is found.

Whilst the *Almagest*, or Sacrobosco's tuned-down version, *De Sphaera*, provided the technical side of cosmology, Genesis delivered the information on the origin. On the first day God created heaven and Earth, and separated light from darkness. On the second day the *firmament* is created. On the fourth day, lights are created to separate day and night. He made two big ones, the bigger one to rule the day, the smaller one to rule the night, and stars He made as well, and placed the lights on the firmament.

The Cosmos also figured prominently in the imagery of the mystics, as shown in a vision of Hildegard von Bingen who lived from 1098 to 1179. Hildegard, abbess of Rupertsberg, was a German mystic whose written work was about religion, medicine, music, and also cosmology. Her treatise *Causae et curae*, which deals with health and illnesses and other aspects of the human condition, begins with a theological–cosmological background of our Universe. She tells about the creation of the world, the construction of the Solar System and the Cosmos, and of the elements that make up the Universe. In the pictorial representation of her visions we see a spherical universe that holds in its centre the spherical Earth, but the true centre is the human being. The Universe is built around Man, the apex of creation.

This was the cosmic order as accepted by practically everyone. Yet, at the beginning of the Renaissance this order was fundamentally questioned. In the middle of the fifteenth century, Cusanus proposed a dramatically different alternative, but had little response. Then, in the middle of the sixteenth century, Copernicus suggested a much less dramatic correction. Finally, in the middle of the seventeenth century, Descartes, supported by Tycho Brahe's and Galileo's observations, reshuffled and enlarged the Copernican Sun-centred universe fundamentally, and this time irrevocably.

Cusanus and his universe without centre or boundary

Cosmology deals with the Universe in its totality. As we have just seen, up to the end of the Middle Ages this consisted of the Earth-centred Solar System, with the stars in a spherical shell placed just beyond Saturn. A change of paradigm was announced in the philosophy of Cardinal Nikolaus von Kues (1401–1464) or, latinised, Nicolaus de Cusa, but better known as Cusanus.

Cusanus lived at a crucial moment in the history of the Church. Under the threat of Turkish invasion of the Byzantine Empire and Constantinople itself, an attempt was made to reunite the Greek and the Roman Church. Constantinople hoped for military assistance, whereas the Roman Catholic Church hoped to

Fig. 2.2 Man and Cosmos in the Middle Ages. Cosmic vision of Hildegard
von Bingen. The Universe, a space built for mankind. In contemplative meditation
it becomes a room for the pictures of the soul or the projections of the psyche.
(*Liber Divinorum Operum*, Lucca, Biblioteca Statale.)

impose its authority and theology on the Eastern Orthodox Church. Cusanus led
a papal delegation, which, in 1437, visited Byzanz to prepare a Council that
should effect a reunion. There he met Bessarion, who would later establish
himself as a dominant figure in the Italian Renaissance. This encounter was
the beginning of a lasting friendship.

Bessarion (1395?–1472) was Greek and was educated in Constantinople. In 1437, the Byzantine emperor made him Bishop of Nicea. He accompanied the Emperor to the Council of Ferrara-Florence in 1438, which should have re-united the Churches. The Council was a success, but the reunion failed all the same. In 1439, Bessarion went to Rome and converted to the Roman Church. The Pope appointed him as Cardinal, and he settled in Italy. Bessarion was a great collector. Cusanus was closely involved in the transfer of Bessarion's exceedingly valuable library of ancient manuscripts to Italy shortly before the fall of Constantinople in 1453. These manuscripts were part of the documentary background of the Renaissance.

Bessarion was very much interested in astronomy and was instrumental in the translation of the *Almagest* from Greek manuscripts into Latin by Georg Peuerbach (1423–1461), a friend of Cusanus. This was intended as a translation with a commentary. However, Peuerbach died prematurely and Johannes Regiomontanus (1436–1476) took over. The work was published in 1496 in Venice with the title *Epytoma in almagestum Ptolomei*. It was from this source that Copernicus learned his Ptolemy.

This was the intellectual environment of Cusanus. His best-known work carries the title 'On Learned Ignorance' (Cusa 1440). The strange title has to do with Cusanus' idea that man can never know everything. If we succeed in knowing what we do not know, then we have attained the level of learned ignorance. From his concept of creative thinking Cusanus derived a hypothesis that has practical importance for us: an exact knowledge of absolute truth is impossible, but there is no limit to approximate truth. Thus, it is an error to consider only those scientific endeavours worth doing – such as mathematics – where the answer represents an absolute truth. We can also strive to approach truth in a step-by-step process (Cusa 1440, Book I). This attitude elevated the social standing of experimental physics and careful astronomical observations, which would later be undertaken by Tycho Brahe and Galileo.

The scholastic astronomy of the Middle Ages taught a finite, spherically shaped universe, in circular motion around the centrally placed Earth. 'On Learned Ignorance', particularly Book II, presents Cusanus' ideas about the Cosmos: the Universe has no boundary and no centre. The Earth does not occupy a central position, nor is it the centre of any heavenly sphere. God is the centre of the Universe, and God is everywhere. The Earth is not at rest. But motion is not defined against an absolute frame of rest, but only relative to other objects. In strong contrast to the Platonic–Aristotelian teaching, Cusanus emphasises that there is no reason why the Earth should be inferior in quality to any other celestial body. He suspects that there are inhabitants in all stellar regions, however, we have no means to know about their nature. He is convinced that the Universe cannot perish. The elements can dissolve into each other, and through the interaction something new can arise, which exists as long as the interaction lasts (Cusa 1440, Book II). Cusanus' reasoning was mainly based on philosophical considerations.

This severe break with medieval scholastic thinking came from a man of the Church, and a cardinal at that. His courage, to differ so drastically from Aristotelian scholasticism, encouraged others to do likewise. Cusanus was well known among scientists and philosophers, such as Giordano Bruno, Copernicus, Kepler, Galileo, Leibnitz, Descartes and Huygens.

A warning by the Church

Giordano Bruno (1548–1600) was born five years after Copernicus had died. He was a kind of spiritual son to Cusanus. For this reason we discuss him here and not after Copernicus, where he would belong chronologically. In contrast to Copernicus, but in agreement with Cusanus, Bruno's universe is infinite; the Sun is a star among innumerable other stars.

Bruno's picture of the Universe was not founded on new observations or a thorough re-evaluation of existing observations, but on speculations that he thought to have derived from first principles. He sought to establish scientifically a continuity among all phenomena of Nature. Between plant, animal and man there is only a gradual but not a qualitative difference. He also created his own pantheistic theology, where, for example, the divine nature of Christ was denied. For these heresies he was tortured and burned alive in the year 1600.

The Ash Wednesday Supper is Bruno's major exposition of how he saw the Copernican system (Bruno 1584). But we should be aware that Bruno was not an astronomer and was not writing an astronomical treatise. The astronomical universe was part of his wider cosmos. Bruno did not add any new knowledge or insight to cosmology. Indeed, in several places it looks as if he might have misunderstood Copernicus. However, his publications certainly helped to spread the Copernican view for which Thomas Digges had already prepared the field (see later).

Bruno probably influenced many of those who came after him; he was sufficiently important to Kepler to figure in his discussion of an infinite universe (Kepler 1610). Others might have deemed it more prudent not to show their acquaintance with the heretic. The Church certainly meant his treatment to serve as a serious warning against the arrogation of the right to independent thinking without the guidance of the Church, and against publishing what conflicted with dogma. Galileo, as well as Descartes, were well aware of Bruno's fate. They either submitted, or published in places beyond the reach of the Holy Office.

Copernicus and the question of an infinite universe

Copernicus (1473–1543) justified his break with the geocentric dogma by a mixture of philosophical and astronomical arguments. They are given in the first of the six books that make up his *De Revolutionibus Orbium Coelestium* or in short *De Revolutionibus* (Copernicus 1543). Except for the Moon, the celestial bodies no longer circle the Earth. And, except for the lunar circle, Copernicus shifted the centres of the planetary orbits from the Earth to the Sun, the new centre of the Universe. Each planetary orbit had its own centre, close to the Sun. To account for deviations from a perfectly regular velocity against the background of the fixed stars, yet trying to save the circular shape of the planetary orbits, their centres did not reside within the Sun. Although Copernicus gave to the Ptolemaic world a new centre, he retained the Platonic–Aristotelian spherical shape, as well as the circular planetary orbits.

The Copernican universe was much more elegant than the classical concept of Ptolemy. For the description of planetary motion, Ptolemy needed two circles for each planet, but Copernicus realised that the Earth–Sun circle could serve as a common denominator for every Earth-planet pair. This was the practical advantage of the heresy. But Copernicus was iconoclastic in two ways. Crystalline spheres guided planetary paths. The basis of this belief dates back to Aristotle. Whether Copernicus really believed in crystalline spheres at all is not clear when reading through *De Revolutionibus*. However, it is clear that his cosmology demanded at least a reshuffling of the spheres if he wanted them to be crystalline. They might be placed around the Sun, but crystalline spheres could certainly no longer be placed around the Earth, or else some would clash with each other.

Copernicus was undecided as to whether the Universe was finite or infinite. In Book I, Chapter 8, he claimed that from a logical point of view a daily rotation of the Earth explains the apparent daily circles of the heavenly bodies more plausibly than a daily rotation of the rest of the Universe around the Earth. He argued that a rotating universe goes against physical intuition: the bigger the Universe and the further away from the centre, the swifter the motion and the stronger the forces that drive the bodies away from the centre of rotation. If there is no limit to size and speed, they grow to infinity. This leads to a paradox:

> according to the familiar axioms of physics, that the infinite cannot be traversed or moved in any way, the heavens will therefore necessarily remain stationary. But beyond the heavens there is said to be no body, no space, no void, absolutely nothing, so that there is nowhere the heavens can go. In that case it is really astonishing if

> something can be held in check by nothing. If the heavens are infinite, however, and finite at their inner concavity only, there will perhaps be more reason to believe that beyond the heavens there is nothing. For every single thing, no matter what size it attains, will be inside them, but the heavens will abide motionless. For the chief contention by which it is sought to prove that the universe is finite is its motion. Let us therefore leave the question whether the universe is finite or infinite to be discussed by the natural philosophers.

However, one is left with the impression that Copernicus saw the realm of the stars as finite, and of course at rest, placed in the celestial shell adjacent to that of Saturn.

The reception for the new cosmological model was mixed. It varied from outright refusal by Philipp Melanchthon, one of the main figures of the German religious Reformation, who cited the Bible, e.g. Joshua 10, 12–13, to prove that the Sun moves and the Earth rests, to wholehearted acceptance by Kepler. *De Revolutionibus* was widely distributed and read. This is proven in Gingerich's very well documented compilation of the early owners of the surviving copies of the first and second editions (Gingerich 2002). It should be emphasised that the book was bought and read by many who refused to accept the heliocentric system. The new cosmology was introduced in Book I; the other five books explained how to calculate the positions of the Sun, the Moon and the planets. Those instructions could be employed without believing in the introductory text.

De Revolutionibus was dedicated to Pope Paul III. The Church voiced no serious objections. Only in spring 1616 did the Holy Office insert *De Revolutionibus* into the index of prohibited books with the proviso 'donec corrigatur' (until corrected), which meant that one was allowed to read it only after ten passages of the text had been corrected. This action was one of the measures taken by the Holy Office in its fight against Galileo Galilei.

Thomas Digges: How far do stars extend?

One of the earliest proponents of the Copernican model was Thomas Digges (*c.* 1546–1595) in England. Since 1553, his father, Leonard Digges (1520–1559), had published several editions of an almanac. Almanacs were very popular in Europe up to the first half of the twentieth century. They contained, among other information useful for daily life, or of cultural interest, weather forecasts, a perpetual calendar and astronomical information.

In 1576, Thomas Digges appended to the *Prognostication everlasting* a description of the Copernican system, entitled 'A perfit description of the Caelestiall

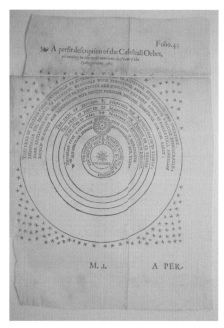

Fig. 2.3 The transition from the geocentric to the heliocentric universe.
Left: The geocentric system. (Leonard Digges, *A prognostication everlastinge corrected and augmented by Thomas Digges*, London, 1576. By permission of the Master and Fellows of Emmanuel College, Cambridge.) Right: The heliocentric system as seen by Thomas Digges. (Given in the Appendix: 'A perfit description of the Caelestiall Orbes'.)

Orbes' (Digges 1576). It is an early example of the switch from the geocentric to the heliocentric concept. Thomas Digges goes a step beyond Copernicus. Whereas Copernicus placed the stars in a shell just beyond that of Saturn, in Thomas Digges' model they fill an unbounded universe. Yet, the heavens of Digges are still of a theological nature. Inscribed in his map of the heavens we find a 'palace of felicity, devoid of grief, and replenished with perfect, endless joy'.

A crucial step in observational techniques

Observational accuracy at the time of Copernicus was approximately the same as at the time of Hipparchus (*c.* 190–125 BC), when stellar positions were given to an accuracy of roughly half a degree. Tycho Brahe (1546–1601) built a new generation of instruments that allowed him an average accuracy of 1 minute of arc, thus better by a factor of 30 (see Maeyama 2002). That accuracy was ideal – not too low, not too high – to lead Kepler to the elliptical shape of the planetary orbits.

Tycho was not happy with the novel arrangement of Copernicus. His argument against the heliocentric system was the lack of stellar parallaxes. If Copernicus were right, the stars nearest to us should show a periodic yearly motion against the background of far-away stars due to our annual path around the Sun. Tycho found no such motion. Copernicus explained this lack of annual parallaxes by the enormous distance of even the nearest stars. Tycho was not prepared to believe in such a huge universe. He opted for an intermediary model, where the Earth remains at rest, the Sun orbits the Earth, the planets orbit the Sun, and the dominion of the stars turns daily around the Earth.

Stellar parallaxes were indeed only found in 1839 by Bessel, Struve and Henderson, the largest being 0.75 seconds of arc (α Centauri). Tycho Brahe's model had a short life, but it gained considerable importance at the time of Galileo, when it was favoured by the Church, for it left the Earth still at its traditional central place.

In spite of Tycho's great achievements in accuracy, no galaxy had yet been discovered in the Western World. That was reserved for the telescope.

Kepler's finite universe and Galileo's telescope

More than sixty years after the first edition of *De Revolutionibus*, the battle for the heliocentric system slowly drifted toward its climax in the open fight of the Church against Galileo Galilei. From the cosmological point of view, it was a fight over the location of the centre of the Universe. The original Ptolemaic, totally Earth-centred model was gradually abandoned when important exponents of the Church, such as Christoph Scheiner (1573–1650), accepted the Tychonian system.

For Kepler there was never any doubt: the Copernican heliocentric system was right. Kepler's universe was finite; a position he strongly defended in the *Dissertatio cum nuncio sideris Galilei*, as well as in his later *Epitome astronomiae Copernicanae* (Kepler 1610, 1618-21). He found observational support in the darkness of the night sky, known today as Olbers' paradox.

In *Dissertatio cum nuncio sideris Galilei*, Kepler refers to Galileo's discovery of the large number of stars unknown before. He is grateful to Galileo for furnishing arguments against Giordano Bruno. 'To avoid that this one [Bruno] converts us to his view of an infinity of worlds, as many as there are fixed stars, all similar to our world, your third observation of the innumerable number of fixed stars beyond those known since antiquity is most helpful.' He argues that the stars must be smaller and of lesser luminosity than the Sun, and their number cannot be infinite: 'if those suns are of the same kind as our Sun, why do they not together surpass our Sun in brightness?' He employs this argument to propagate the idea that the Sun has unique qualities, and that the stellar shell surrounding the Sun cannot be of infinite extent.

In his *Epitome astronomiae Copernicanae*, Kepler outlines his ideas about the structure and dynamics of the Universe. We summarise some points from Book IV, Part II. Tycho Brahe had shown that there are no material spheres guiding the planets around the Sun. What force then guides the planets? Kepler did not want to fall back on the supernatural; he looked for guiding forces that were open to scientific investigation. In his opinion the Sun is the central and most important body of the Universe. It rotates around its own axis in the same sense as the planets turn around the Sun. It is endowed with a force that makes the planets move: 'Because it is apparent that in so far as any planet is more distant from the sun than the rest, it moves the more slowly … Therefore we reason from this that the sun is the source of the movement.' (Kepler 1618–1621, p. 55.) What qualities does this driving force possess?

Kepler mentions on several occasions the action of the lodestone. In 1600, Gilbert had described magnetism scientifically in his publication, *De Magnete*. A magnet can act without touching the other body, why should similar forces not be active between the Sun and the planets! When two lodestones meet, their 'friendly' parts attract each other, whereas the opposite parts are 'unfriendly'. It might be similar for the Sun and the planets: 'Just as there are two bodies, the mover and the moved, so there are also two powers, by which the movement is administered; one is passive and verges more towards matter, namely the likeness of the planetary body to the Solar body, with respect to the bodily form; and there is one part of the planetary body which is friendly to the Sun, and the opposite part is unfriendly. The other power is active and smells more of form – that is to say, the Solar body has the force to attract the planet with respect to its friendly part and to repulse it with respect to its unfriendly part, and finally to keep it, if it were placed thus, so that it does not direct either its "friendly" or its "unfriendly" part against the Sun.' (Kepler 1618–1621, p. 57.)

The primary mover was no longer hiding in the outermost sphere, but had become a physical force, located in the Sun. Although Kepler got it wrong, it was a first step to astrophysics: the motion of celestial bodies had been assigned a natural cause.

For Kepler, the Sun and its planets stood in the middle of the Universe. In Part I of Book IV, Kepler reasons that, based on Tycho Brahe's observations and his own speculations, the distance to the sphere of the fixed stars should be about 2000 times the distance of Saturn. (He thus underestimated that ratio by more than a factor of ten.) That spherical shell contains a large, but finite number of stars, and it is at rest. This is also plausible from an extrapolation of the planetary orbital periods. The ratio of their periods increases with the power 3/2 of their distances from the Sun. As the period of Saturn is already 30 years, the stellar periods would be several million years.

Not only Kepler, but Halley, Cheseaux (1718–1751), Olbers (1758–1840) and several others argued that if the Universe is infinite and evenly filled with stars, we would see one of them in whatever direction we turn our eyes, and thus, if stars are of the same nature as our Sun, the sky would be bright day and night. This reasoning rested on the assumption of a geometrically flat Euclidean universe, with a beginning far remote in time, without interstellar absorption, and with stars that would live forever. Whereas Cheseaux (1744, p. 225) and Olbers (1823, p. 110–121) suspected some minuscule interstellar absorption to extinguish remote starlight, Olbers' paradox helped Kepler to defend his conviction that the Universe was finite.

An infinite universe would have been contrary to Kepler's philosophy, embodied in his geometrical construct, and beautifully presented in the *Mysterium Cosmographicum* (Kepler 1596). The perfection of the Universe resides in its measured harmonies within a finite volume; infinity would be alien to such a concept.

While Kepler worked out the laws describing planetary orbits, in 1609 Galileo Galilei (1564–1642) inaugurated a new epoch in the exploration of the heavens with his telescope. New stars and heavenly phenomena that no human eye had seen before became visible: moons around Jupiter, mountains on the Moon, and a much larger number of stars than had ever been counted before. The Universe obviously contained much more than had hitherto been known.

In the *Sidereus Nuncius*, where Galileo announced his discoveries, he tells us that he had learnt about an optical instrument that brought remote objects much closer (Galileo Galilei 1610). He then built a telescope himself and followed up his discoveries with a well-planned and highly successful publicity campaign. He was 46 years old and wanted to be famous in order to obtain a better paid position. The wonders seen were indeed an irresistible incentive to improve that optical instrument. Even Newton took part in this endeavour and invented the reflecting telescope. The efforts paid handsome dividends.

In his writings Galileo did not present a new grand-scale picture of the Universe. It would anyway have been a dangerous venture. His summons to Rome in 1632 reminded him that in 1600 Giordano Bruno had paid with his life for heretical views. The Church still had the power to deal mercilessly with unorthodox opinions.

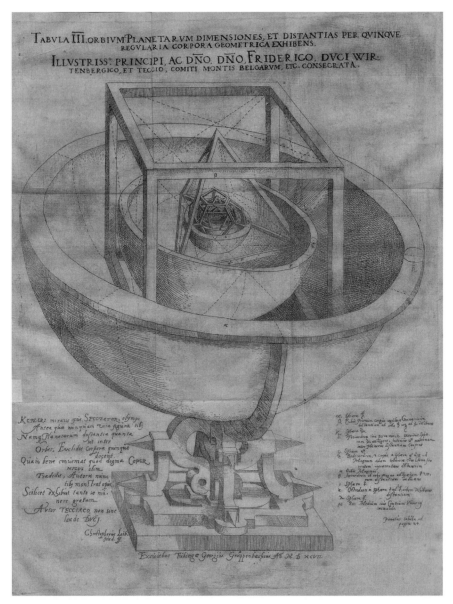

Fig. 2.4 Kepler's universe. The Sun is in the centre. The planets move in individual shells. Distances between the shells are determined through the Platonic bodies (tetrahedron, hexahedron [cube], octahedron, dodecahedron, icosahedron). The outer boundary of the planetary shell lies as closely as possible around the body, and the inner boundary fits as large as possible within the body. This universe is finite. When Kepler realised that the accuracy of the predicted planetary orbital sizes was deficient, he no longer used the model for detailed calculations, but he retained the principle of a geometrically constructed finite universe. (Johannes Kepler, *Mysterium Cosmographicum*, 1596. ETH-Bibliothek, Zurich, Sammlung Alte Drucke.)

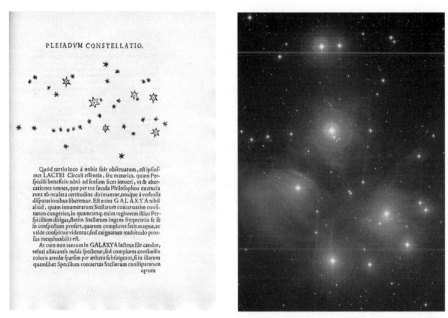

Fig. 2.5 Extending the heavens. Left: A page from *Sidereus Nuncius*, where in 1610 Galileo Galilei showed the known and the newly discovered stars in the Pleiades. (Galileo Galilei, *Sidereus Nuncius*, 1610. ETH-Bibliothek, Zurich, Sammlung Alte Drucke.) Right: The Pleiades, seen by modern telescopes. Just turn this picture by 45 degrees to appreciate the accuracy of Galileo's drawing. (NASA/Caltech, Palomar 48-inch Schmidt Telescope.)

Descartes: An evolving universe

In contrast to Digges' design, there was no room for a theological heaven in the boundless universe of Descartes (1596–1650). In 1644, thus one hundred years after Copernicus' *De Revolutionibus*, Descartes published in *Principia philosophiae* his description of a universe without a centre and without limits. For theological reasons a distinction was made between an infinite universe and a universe without limits; the quality of true infinity was left to God. Anyway, true infinity could not be experimentally proven. In the first part of his Principles (Descartes 1668, Section 26), Descartes said about the infinite: 'Qu'il ne faut point tâcher de comprendre l'infini mais seulement penser que tout ce en quoi nous ne trouvons aucunes bornes est indéfini.' In Section 27 he continued: 'Et nous apellerons ces choses indéfinies plutôt qu'infinies, afin de réserver à Dieu seul le nom d'infini.' (We should not try to understand infinity, but consider anything as indefinite, where we find no boundary. And we call these things indefinite rather than infinite, to reserve for God alone the name of the infinite.)

Descartes cleared the way for a modern view of the Universe. Even before the Copernican ideas were generally accepted, Descartes declared them as not going sufficiently far, and he dislodged the Sun and the Solar System from their privileged central position. Following in the footsteps of Cusanus and Giordano Bruno, and contrary to the opinion of Kepler, Descartes declared the Sun to be a star among others, and advocated a universe that knew no boundary.

Descartes is one of the founders of modern mathematics and physics. His cosmological theories are based on metaphysical reflections rather than on observation. However, he did employ observational arguments to distinguish between the illuminated planets and fixed stars, which shine with their own light and are of the same nature as our Sun. His view of the world was strictly mechanical. In the third chapter of his *Principia philosophiae* he sketched a cosmological history. God created the substance that constitutes the Universe; after his initial impetus the Universe now runs according to natural laws. These laws are accessible to the scrutiny of our brain.

Descartes' contemporary, Blaise Pascal (1623–1662), himself an outstanding scientist, did not appreciate this removal of God to the status of a passive onlooker at the fringe of the Universe. Descartes and Pascal represented two incompatible tendencies. Descartes strongly supported the separation of religion and science. Pascal hoped to merge them both on an equal footing, but in the end acknowledged the supremacy of religious belief. Contrary to Newton's credo of an intervening God, who from time to time had to mend his creation to prevent collapse, Descartes had no room for a God of miracles.

What then is the force that keeps the planets on their path? Can a force act through empty space? With his hypothesis of vortices Descartes had an answer to the question of action at a distance. Vortices are the basic elements of his universe, serving three purposes: they guarantee that there is no empty space – space without content does not exist – and they create and hold together the heavenly bodies. They also structure and maintain the Universe in a process of perpetual dynamics. His cosmological events are the result of a kind of fluid dynamics.

Descartes' most revolutionary aspect is his introduction of evolution into cosmology. He refused the biblical concept of the original creation as given in Genesis. Our Universe is the result of evolution. His vortices created planets and stars. No god is needed; Nature does it all. This concept, that matter possesses an inherently creative potential, will later be pursued explicitly by Kant and Laplace, and we will find it implicitly when Herschel interprets his observations of nebulae. Descartes also had a considerable direct influence on Newton, not only on his mathematics but also on his conception of the Universe – although as an outspoken antipole in the case of vortices. In the next section we will come back to Newton's very outspoken opposition to Descartes' concept of evolution.

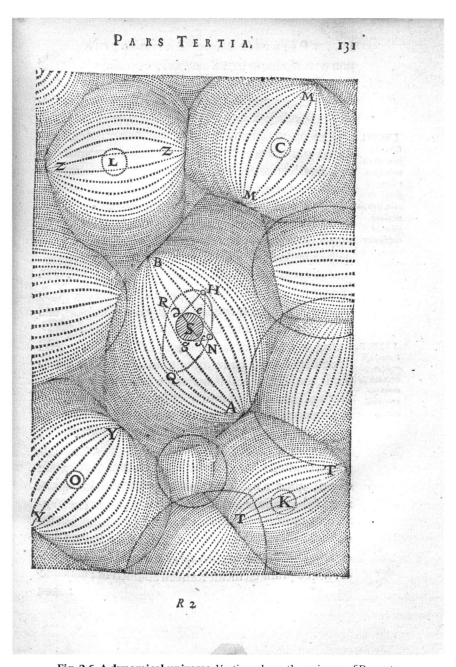

Fig. 2.6 A dynamical universe. Vortices shape the universe of Descartes. (René Descartes, *Principia philosophiae*, 1644. ETH-Bibliothek, Zurich, Sammlung Alte Drucke.)

Descartes' cosmological model – a universe structured by vortices – had to be dropped by the end of the seventeenth century. With Newton's *Principia*, gravitation became accepted as the dominant force in the Universe. Descartes' vortices had given a qualitative description, but, in contrast to Newtonian theory, they had no precise predictive power, and Newton showed that their action within the planetary system would be incompatible with Kepler's laws (Newton 1726, Book 2.9).

Descartes' lasting contributions were his analytical geometry and his introduction of evolution into cosmology. But Descartes was on the wrong track when trying to find the forces that structured the Universe. Newton and his circle would tackle this problem half a century later.

Newton's *Principia* and Bentley's sermon

Newton's *Principia*, first published in 1687, provides a single set of laws that govern the motions on Earth and in the heavens. Gravitational attraction and Newton's laws now explain the action of the tides, the precession of equinoxes, the orbits of the planets and their moons, and even the paths of comets. They became the physical foundation of the emerging new astronomy. The difference in emphasis between Newton and Descartes is shown in the shift of the title from Descartes' *Principia philosophiae* to Newton's *Philosophiae Naturalis Principia Mathematica*. Essentially, Newton did not venture beyond what could be established empirically. We will see that this does not apply to the 'General Scholium', but that was really written as an appendix.

Descartes' vortices were the means to pass an impetus from one body to the next by direct contact. Gravitation does not act by direct contact; it is a long-range action, a force at a distance. Newton postulated that gravitation exists in all bodies universally and is proportional to the quantity of matter in each; it is very likely that Hooke proposed this idea originally (Hooke (1674), last part of his Cutler lecture). In that respect, the force of gravitation is different from the magnetic force that Kepler considered a candidate for moving the planets. Another difference between gravitation and magnetism remarked upon by Newton: the magnetic force decreases not as the square, like gravitation, but as the cube of the distance. But what is this gravitation that rules the motion of all the bodies in the Universe? Newton admits that he has 'explained the phenomena of the heavens and of our sea by the force of gravity, but I have not yet assigned a cause to gravity … I have not as yet been able to deduce from phenomena the reason for these properties of gravity, and I do not feign hypotheses [Hypotheses non fingo].' (Newton 1726, 'General Scholium'.)

Although the third book of the *Principia* carries the title 'The System of the World', Newton did not really propose a new structure of the Universe, as had been done by Copernicus or Descartes. But in his 'General Scholium', which was added in the second edition of 1713, he speculated on the grand design. Newton accepted the Cartesian concept of a world that may have no boundary, and of stars being similar to our own. But he refused Descartes' vortices as the agents structuring the Universe; gravitation rules the world.

Planets and comets all move according to the laws of gravitation, but they could not originally have acquired the regular positions of the orbits by these laws. 'This most elegant system of the sun, planets, and comets could not have arisen without the design and dominion of an intelligent and powerful being. And if the fixed stars are the centres of similar systems, they will all be constructed according to a similar design and subject to the dominion of One, ... And so that the systems of the fixed stars will not fall upon one another as a result of their gravity, he has placed them at immense distances from one another.' (Newton 1726, 'General Scholium'.)

Some of Newton's ideas on cosmogony are contained in his letters to the Reverend Dr Richard Bentley, chaplain to the Bishop of Worcester. We recover some of these ideas in Bentley's famous sermons 'A confutation of Atheism', preached at St Martin's in the Fields in the late autumn of 1692 and published in 1693. Reading Bentley's sermons you are left with the strong impression that the two men must have synchronised their views before the sermons were delivered.

In his first letter, Newton outlines, but then rejects, a possible way of forming stars and planets. He agrees that matter scattered throughout a finite space would collapse into one great spherical mass. 'But if the matter was evenly dispersed throughout an infinite space, it would never convene into one mass, but some of it would convene into one mass and some into another, so as to make an infinite number of great masses, scattered at great distance from one to another throughout all that infinite space.' However, he cannot see how in this way matter could divide itself into two sorts, the one for the shining bodies, like the Sun, and the other fit to form planets. He does not think this separation explicable by mere natural causes, but feels forced 'to ascribe it to the counsel and contrivance of a voluntary agent' (Newton 1692, p. 94).

In Bentley's crusade of 1692 against the 'illiterate and puny Atheists', the sermons are peppered with scientific arguments. He intended to prove that God was needed to create the present harmonious universe, and that it needs God to maintain it in a stable state. In this spirit he lectured about planetary orbits: 'though we grant, that these Circular Revolutions could be naturally attained; or, if they will, that this very individual World in its present posture and motion was actually formed out of Chaos by Mechanical Causes: yet it requires a Divine Power and Providence to have conserved it so long in the present state and condition.' (Bentley 1693.) And he continued with brilliant rhetoric: 'Now what Natural Cause can overcome Nature itself?' The stars with their systems of planets would by the action of gravitation all have collapsed toward the most central system in the Universe. 'It is evident therefore that the present Frame of Sun and Fixt Starrs could not possibly subsist without the Providence of that almighty Deity, ...' Bentley gives no definite answer whether the Universe is finite or infinite, but he tends towards a finite version.

We may be certain that the theologian's sermon on the inevitability of a universal gravitational collapse, unless God intervened, mirrored Newton's conviction. Had Newton been less categorical in his refusal of vortices, he might have discovered the stabilising effect of random motion that prevents a gas cloud, as well as a globular cluster, from unrestrained gravitational collapse. Under the sole influence of gravitation a static universe is doomed to some kind of collapse. Newton knows of no remedy – except for divine intervention – and does not contradict Bentley.

Newton was firmly opposed to Descartes' concept of evolution. This is expressed in the famous Query 31 at the end of his *Opticks*: 'all material Things seem to have been composed of the hard and solid Particles above-mention'd, variously associated in the first Creation by the Counsel of an intelligent Agent. For it became him who created them to set them in order. And if he did so, it's unphilosophical to seek for any other Origin of the World, or to pretend that it might arise out of Chaos by the mere Laws of nature; ... For while Comets move in very excentrick Orbs in all manner of Positions, blind Fate could never make all the Planets move one and the same way in Orbs concentrick ...' (Newton 1730, p. 402.)

Thus, the cosmology of the beginning of the eighteenth century inherited from the seventeenth century Descartes' philosophical concept of an evolving, unbounded universe. With Newton's laws that govern the motion of the bodies in that universe, it now possessed a most powerful scientific theory. Nevertheless, religious traditions lingered on and strongly influenced scientific thinking.

3
_____

Nebulae as a new astronomical phenomenon

The Sun, the Moon and the stars are easily recognised objects, what about nebulae? The first written Western report about a nebula dates from 1614 and is from Simon Marius. Seventy years later, the Andromeda nebula appeared prominently in the beautiful catalogue of Hevelius, printed in 1687. However, nebulae only became a respected astronomical topic when, in 1716, Halley highlighted them in the *Philosophical Transactions*. He introduced them with a strong biblical colouring. That religious connotation became dominant in Wright's divine universe of 1750. However, Immanuel Kant would have none of it and led cosmology through the sharpest possible U-turn on to a scientific path. A great admirer of Newton's *Principia*, he proposed in 1755 what became known as the theory of island universes. Forty years later, without knowing the work of Kant, Laplace proposed a similar universe. That set the stage for Herschel.

Early reports on nebulae

To our knowledge, the first description of a nebula in Western astronomy is from Simon Marius (1573–1624) concerning the great nebula in Andromeda. In the summer of 1608 he learnt of a new instrument, built by a Belgian then living in France. It allowed him to see distant objects as if they were very close. From the summer of 1609 onward, Marius was able to observe the sky with such a Belgian telescope, which belonged to his benefactor. With this instrument he discovered the moons of Jupiter at about the same time as Galileo, but he only published his observations in 1614 in his book, *Mundus Iovialis* (Marius 1614). In that same publication he reported that, since 15 December 1612, he had observed close to the northerly third star in the belt of Andromeda a star of astonishing shape, unlike any he could find in other parts of the sky. Without an instrument

Fig. 3.1 The Andromeda nebula. Al Sufi (903–986) published an update of Ptolemy's catalogue of fixed stars. In the constellation of Andromeda he entered a 'cloudlike spot', which in this parchment is given as an assembly of points in front of the fish's mouth. It is the spiral galaxy M31. This picture, now at the Universitäts- und Forschungsbibliothek Erfurt/Gotha, is from an Italian parchment of al Sufi's *Book of Fixed Stars*, dated 1428. (Gotthard Strohmaier, *Die Sterne des Abd ar-Rahman as-Sufi*, Gustav Kiepenheuer Verlag, Leipzig und Weimar © Aufbau Verlagsgruppe GmbH, 1984.)

he sees there something like a nebulosity. However, when looking through the telescope he could not discern individual stars, as in some stellar clouds, but only glittering rays, which grew brighter the closer they were to the centre. In the centre, there was a feeble and pale radiance with a diameter of approximately a quarter of a degree. A similar radiance could be found when observing a burning candle through a translucent horn from a great distance. Marius further remarks that he does not know whether the star is a new one. He is astonished that the sharp-eyed Tycho did not notice this nebula, although he determined the coordinates of a star just a little north. There can be no doubt that Simon Marius had observed the great nebula in Andromeda. It became object number 31 in Messier's catalogue of 1771, which we designate as M31.

The oldest pictorial presentation of M31 is probably the one by Abd al-Rahman al Sufi (903–986), the tenth-century Persian astronomer. He had a fresh look at Ptolemy's catalogue of fixed stars, which he enriched, corrected and updated. He also combined Arabic star names with the Greek constellations. His illustrated *Book of Fixed Stars* was copied many times. From a 'westernised' edition, dated 1428, produced most probably in northern Italy, we show an illustration of the Andromeda constellation, featuring the nebula M31 (see Strohmaier 1984).

Edmond Halley on several nebulae or lucid spots like clouds

At the end of 1715, in No. 346 of the *Philosophical Transactions of the Royal Society of London*, an account was published about several 'new stars' that had appeared within the previous 150 years – to us known as novae and supernovae. In No. 347 there followed 'An Account of several Nebulae or lucid Spots like Clouds, lately discovered among the Fixt Stars by help of the Telescope'. No author is given for either of the two publications. However, Halley was at that time secretary of the Royal Society and, in 1733, Derham – whom we shall soon meet – attributed the authorship of the 'Account on Nebulae' to Halley; we therefore assume that Halley was indeed the author. Referring to the new stars about which he had reported before and to the Book of Genesis, Halley writes (Halley 1716):

> But not less wonderful are certain luminous Spots or Patches, which discover themselves only by the Telescope, and appear to the naked Eye like small Fixt Stars; but in reality are nothing else but the Light coming from an extraordinary great Space in the Ether; through which a lucid Medium is diffused, that shines with its own proper Lustre. This seems fully to reconcile that Difficulty which some have moved against the

Description Moses gives of the Creation, alledging that Light could not be created without the sun. But in the following Instances the contrary is manifest: for some of these bright Spots discover no sign of a Star in the middle of them; and the irregular Form of those that have, shews them not to proceed from the Illumination of a Central Body.

Well, on the first day God said: Let there be light; and there was light. However, the Sun was made on day four only. But the discovery of nebulae, shining with their own proper lustre, restored peace to the doubtful mind, showing that light could well have existed before the Sun was made. Halley was visibly relieved to have recaptured that trump card out of the hands of the 'illiterate and puny Atheists' slated twenty years earlier by Bentley.

Before entering the subject of nebulae, let us very briefly summarise what we know today about them. There are extragalactic and galactic nebulae. The extragalactic nebulae are galaxies in their own right. As they lie far outside our own galaxy – the Milky Way – they usually cannot be resolved into individual stars, except for the most luminous stars in nearby galaxies. Once these nebulae were recognised as galaxies in their own right, they began to be seen as building blocks of the Universe. Nebulae within our own Galaxy are much smaller objects (Galaxy written with a capital stands for our own 'island universe', the Milky Way). They are either dense stellar agglomerations containing up to several hundred thousand stars, such as globular clusters, or they are objects with dominant gaseous ingredients, such as planetary nebulae or interstellar gas- and dust-clouds like the Orion nebula. Orion is a bright nebula where very hot young stars illuminate their gaseous environment. Orion is the first nebula on Halley's list.

The great nebula in Andromeda is the second entry on Halley's list, and is also known as M31. It is an extragalactic nebula of roughly the same size as our Galaxy. Halley points out that 'it is inserted into the Catalogue of Hevelius, who has improperly called it *Nebulosa* instead of *Nebula*; it has no sign of a Star in it. But appears like a pale Cloud …' Indeed, in his *Catalogus stellarum fixarum* of 1687, Hevelius identified it as 'Nebulosa Cinguli. In Crinibus sub Brachio dextro' (Hevelius 1687). (Nebulosity in the girdle. In the hair below the right arm.) We show here the Andromeda constellation in Hevelius' posthumously published and beautiful celestial atlas of 1690 (Hevelius 1690).

Halley saw that these nebulae remained fixed among the fixed stars and had therefore to be very far away. He further observed that, contrary to the fixed stars, they were seen as extended objects, so they must therefore have huge diameters. Halley ends his report: 'In all these so vast Spaces it should seem that there is a perpetual uninterrupted Day, which may furnish Matter of

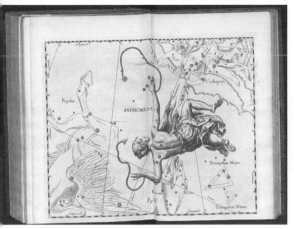
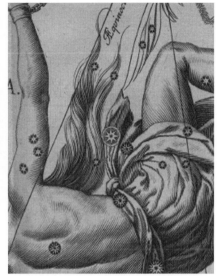

Fig. 3.2 Andromeda by Hevelius. Left: The constellation of Andromeda. Right: Detail with the spiral galaxy M31 in the very centre, represented as a large encircled star. Hevelius drew his constellations as seen from the outside of the heavenly sphere. (Johannis Hevelius, *Firmamentum Sobiescanum, Uranographia*, 1690. ETH-Bibliothek, Zurich, Sammlung Alte Drucke.)

Speculation, as well to the curious Naturalist as to the Astronomer.' That it did indeed.

In the *Philosophical Transactions* of 1733 we find an article by W. Derham, Canon of Windsor, with the title 'Observations of the Appearances among the Fix'd Stars, called Nebulous Stars' (Derham 1733). He speculates about the nature of these nebulae: 'I conclude them certainly not to be *Lucid Bodies*, that send their Light to us, as the Sun and Moon. Neither are they the *combined Light of Clusters* of Stars, like the Milky-Way: But I take them to be *vast Areae*, or *Regions of Light*, infallibly *beyond the Fix'd Stars, and devoid of them*.' (Italic writing by Derham.) And he further speculates about there being openings into an immense region of light, beyond the fixed stars.

The universes of Wright, Kant and Laplace

The religious associations of Halley and Derham were far outdone by the cosmic vision of Thomas Wright (1711–1786). He proposed several versions. In principle, the stars are arranged either in a spherical shell or in an annulus, separated from a supernatural centre by a huge gap. The stars are in motion about this centre. In analogy to the planetary orbits in the Solar System, circular

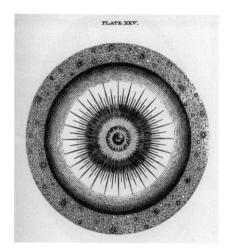
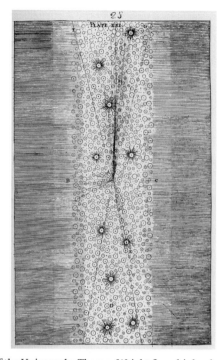

Fig. 3.3 Thomas Wright. A model of the Universe by Thomas Wright from his book *An original theory or new Hypothesis of the universe*, published in 1750. Left: A central Section of the Convexity of the entire Creation, with the Eye of Providence seated in the Center. Right: A small portion of the shell of stars with the observer in the centre, who sees many stars when looking along the Milky Way, but only a few in the perpendicular direction. (Facsimile of the 1750 edition with an introduction by M. A. Hoskin, London: Macdonald, 1971.)

motion prevents the cosmic system from collapsing. Wright gave a detailed model of the Milky Way, combining theological and astronomical concepts. The stars are of the same nature as our Sun. They are grouped in a spherical shell around the divine centre. He leaves the possibility open that there is more than one shell. The shell is vast, containing innumerable stars. The distribution of stars is not absolutely even, which explains why in some directions we see larger groups than in others. To an observer who lives within the shell – such as observers on Earth – the stars appear to lie in a plane: the tangential plane to the sphere at the point where we are located. The stars move, and Wright indicates several possibilities of how they might be moving. As mentioned before, the centre of his universe is of divine nature, there the 'divine Presence, or some corporeal Agent, full of Virtues and Perfections, more immediately presides over his own Creation' (Wright 1750). Wright's cosmology was really meant to fuse astronomy and theology.

Wright's confused and confusing writing would soon have been forgotten, had it not been for Immanuel Kant (1724–1804). He read about Wright's model. It inspired some of the cosmological concepts in his *Allgemeine Naturgeschichte und Theorie des Himmels* (Universal Natural History and Theory of Heaven), published in 1755. In the extension of the title, Kant explains that he presents an attempt to treat the origin and the present state of the whole universe (Weltgebäude) according to Newtonian principles. He acknowledges his debt to Wright, of whose publication he learnt from a review in the *Hamburgischen freien Urtheilen* of 1751. It made him see the fixed stars not as an unordered crowd, but ordered, lying in a plane, with their motions governed by Newton's laws. The Milky Way consists of innumerable stars that are crowding along that plane, producing the nebulous appearance; our Sun lies close to this plane. This is his debt to Thomas Wright, whose further divine speculations he gently ridicules.

Kant discusses earlier astronomical publications. He cites Bradley in support of his view that all the stars are in motion, and he mentions Halley for the list of nebulae published in the *Philosophical Transactions*. This leads to his hypothesis of the 'island universes'. He refers to Maupertuis, in whose work he had seen nebulae represented as elliptically shaped patches. He now combines that information with his Wright-inspired picture of the Milky Way, as a disk-shaped agglomeration of stars. If such a 'World of Fixed Stars' is seen from a very great distance, it will appear as a very small patch, dimly lit, and of elliptical shape when not seen face on, where it would appear circularly shaped. These elliptical figures have to be seen as World Systems, just like our Milky Way (Kant 1755). We have here the first suggestion of a universe composed of individual galaxies, one of them being our Milky Way. That universe has no boundary. In the second part of his 'Theory of Heaven' he strongly advocates that our present universe is the result of an evolution.

Kant's book did not become known until the middle of the nineteenth century. It seems that bankruptcy of the publisher stopped the distribution. A wider impact was reserved for Pierre-Simon Laplace (1749–1827) with his *Exposition du système du monde*, first published in 1796. It was written as a non-mathematical introduction to his most important work, *Traité de Mécanique Céleste*, published in 1799. The *Exposition*, which had a wide readership, consists of five books. The first is about the apparent motions of the celestial bodies, the second about the real motions, the third about the laws of motion and the fourth about the theory of gravitation. The title of the fifth announces the history of astronomy, but it also contains his nebular hypothesis and his well-known remark about black holes: 'Un astre lumineux de même densité que la terre, et dont le diamètre seroit deux cent cinquante fois plus grand que celui du soleil, ne laisseroit en vertu de son attraction, parvenir aucun de ses rayons jusqu'à nous; il est donc possible que les plus grands corps lumineux de l'univers, soient par cela même, invisibles.' (A luminous star of the same density as the Earth, the diameter of which is 250 times that of the Sun, would, because of its attraction, not allow any of its rays to reach us; it is therefore possible that the largest luminous bodies of the Universe are for this reason invisible to us.)

Laplace postulated that the Solar System originated from a large, flattened, slowly rotating, contracting cloud of gas, and he gives detailed justifications for that assumption. Then he directs his attention beyond the Sun, where innumerable stars may well be centres of additional planetary systems.

Laplace then gives his view on nebulae and cosmology. Stars seem not to be distributed evenly throughout the Universe. They are assembled in groups, each of them containing several billions (plusieurs milliards) of stars. Our Sun and the brightest stars around us are probably, he writes, members of such a group that forms the Milky Way. Seen from a sufficiently large distance, the Milky Way would appear just like one of those nebulae that, due to their enormous distances, cannot be resolved into individual stars (Laplace 1796). Laplace formulated his theoretical concepts contemporaneously with Herschel's nebular observations.

Although there are significant differences of detail concerning the structure of the Universe as an unbounded conglomeration of individual galaxies, as well as the birth of stars and planetary systems out of rotating clouds of gas, Laplace arrived at similar conclusions to those of Kant before him, even though he most probably did not know about Kant's *Theory of the Universe*. The high scientific reputation and literary style of the famous mathematician assured the *Exposition* an important place among the writings that shaped the opinion of scientists and the lay public.

With Descartes, Kant and Laplace, the qualitatively new concept of evolution entered the scientific discussion of the Universe. They also symbolise the gradual liberation of scientific thinking from the chains of religious dogma. The Christian doctrine of the creation of the Universe by an act of God was deeply entrenched – a God who created heaven and Earth and everything in it, as we know them today, within six days. Descartes, Kant and Laplace replace this view by an evolving universe. Soon Charles Darwin would replace Adam and Eve by the theory of evolution of all things living, including mankind. Whereas Descartes and Kant still accepted God as the creator of matter, it is reported that Laplace's answer to Napoleon's question why he did not mention God, the creator, was: 'Sire, je n'avais pas besoin de cette hypothèse-là.' (Sir, I had no need for such a hypothesis.)

4

On the construction of the heavens

When Galileo Galilei pointed his telescope to the sky, he found assemblies of individual stars, where the naked eye had seen only a hazy nebula. Yet Marius was unable to resolve Andromeda into individual stars. Was it too far away to allow such resolution, or was it composed of truly nebular matter? The philosopher Kant and the mathematician Laplace thought nebulae to be stellar agglomerates and elevated them to building blocks of the Universe.

Observational investigations into their true nature began with William Herschel's ambitious endeavour, 'On the Construction of the Heavens', as the eighteenth century turned to the nineteenth. With their giant telescopes, Herschel and William Parsons (Earl of Rosse) inaugurated a new era in astronomical observation, driven by the desire to solve the nebular enigma. A further qualitative step in nebular research was Huggins' introduction of spectroscopy. We shall follow Herschel's vacillating opinion about nebulae, Rosse's discovery of spiral structure, and Huggin's demonstration that nebulae did not form a unique physical class of objects. With Scheiner's photographic spectrum of M31 (Andromeda) we touch the twentieth century.

Herschel confirms and then rejects island universes

In 1784, barely seventy years after Halley's publication on nebulae, William Herschel turned over a new leaf of nebular research. In a series of publications in the *Philosophical Transactions* he greatly surpassed in number and quality what had been observed before. He announced that his new Newtonian reflector of 20 feet focal length and $18\frac{7}{10}$ inches aperture enabled the resolution of nebulae into individual stars in several cases where the nebulae had not revealed their nature to Messier and Mechain, Messier's younger friend and competitor in the discovery of new comets. He proudly added: 'I have already found 466 new nebulae and clusters of stars, none of which, to my present knowledge, have been seen before by any person; most of them, indeed, are not within the reach of the best common telescopes now in use.' (Herschel 1784.)

Messier had started his nebular catalogue in 1764. He was an eager comet-observer, and catalogued nebulae to avoid being confounded in his search for comets. When first published in 1771, his catalogue contained 45 nebulae (Messier 1774). It was further extended and Messier's final version of 1781 contained 103 entries. Each entry was provided with a comment about special characteristics, e.g. 'it contains no star'. He also mentioned all the other observers who, to his knowledge, had seen the nebula before him. However, Messier was hunting for comets and showed no real interest in the nature of nebulae.

When Herschel turned his full attention to nebulae he was already a seasoned observer. His discovery of Uranus in 1781 had established his worldwide fame. It also procured him a lifelong pension from King George III, which allowed him to make astronomy his main occupation. He had discovered Uranus with a mirror of 15 cm diameter, whereas for his nebular investigation he now had a 47.5-cm mirror with a 6.1-m focal length. Herschel was determined to fathom the nature of nebula and to solve an enigma that had puzzled astronomers since Marius' publication of 1614 and Halley's discussion in 1716.

Herschel was overwhelmed by the number of nebulae he found. The title of his long article of 1785, 'On the Construction of the Heavens', shows that he was aiming high. One of his first statements was: 'That the milky way is a most extensive stratum of stars of various sizes admits no longer of the least doubt; and that our sun is actually one of the heavenly bodies belonging to it is as evident.' But he wanted to go further: 'in order to develop the ideas of the universe, that have been suggested by my late observations, it will be best to take the subject from a point of view at a considerable distance both of space and of time.' (Herschel 1785.)

At the beginning of that same publication he also confessed part of his scientific credo: 'If we would hope to make any progress ... we ought to avoid two opposite extremes ... If we indulge a fanciful imagination and build worlds of our own, we must not wonder at our going wide from the path of truth and nature; but these will vanish like the Cartesian vortices, that soon gave way when better theories were offered. On the other hand, if we add observation to observation, without attempting to draw not only certain conclusions, but also conjectural views from them, we offend against the very end for which only observations ought to be made. I will endeavour to keep a proper medium; but if I should deviate from that, I could wish not to fall into the latter error.'

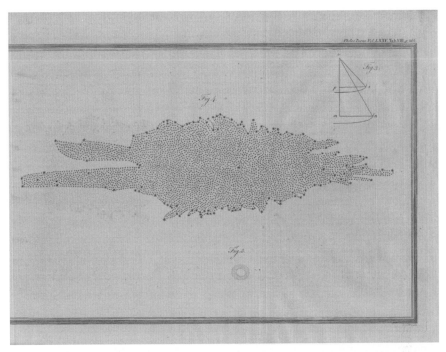

Fig. 4.1. Herschel's image of the Milky Way. Herschel obtained this map by statistical methods, counting stars and assuming that they were evenly distributed within the Milky Way. The large star close to the centre represents the place of the Sun. This picture shows the bulk of the Milky Way with dimensions of approximately 620 times the distance to Sirius in length and 200 Sirius distances in width. This model of the Milky Way was considered qualitatively valid until, in 1918, Shapley found our place to be rather at the outer fringes. (Herschel 1785.)

In 1785, he was full of enthusiasm for his extended nebulae, which 'may well outvie our milky-way in grandeur' (Herschel 1785, p. 260). To explain the observed distribution of stars and stellar clusters, he assumed that from an originally even distribution of stars, clusters had formed due to gravitation. As a natural consequence, large gatherings of stars exist, with great cavities between them. By counting stars along a line of sight and relating that to their magnitudes, he reached the conclusion that we live in a detached nebula, and that 'there is but little room to expect a connection between our nebula and any of the neighbouring ones' (p. 248).

He then set himself the task of delineating the extent of our own galaxy: the Milky Way. His observations suggested that the Sun is part of it and situated close to the centre. He very much underestimated its size. For the bulk of the Milky Way he gave a length of approximately 850 times the distance to Sirius – which is separated from the Earth by 2.6 pc – and a width of approximately 200 times the distance to Sirius.

Herschel considered that the process of agglomeration to denser clusters and the division of larger ones into units was still in progress. He was convinced that some of his newly discovered nebulae could well surpass our own Milky Way in size. The nebula in Andromeda was of the same kind, and he credited it with a distance not exceeding 2000 times our distance to Sirius. But he also realised that some nebulae may be of a different nature from others, some are milky ways like ours, others form part of our own Milky Way, and there might be still others, such as planetary nebulae: 'their singular appearance leave me almost in doubt where to class them.'

Herschel followed this up with his 'Catalogue of One Thousand new Nebulae and Clusters of Stars'. His observations were leading him to the island universes of Kant and Laplace. In the introduction, he reiterates his belief in the existence of isolated stellar systems: 'to the inhabitants of the nebulae of the present catalogue, our sidereal system must appear either as a small nebulous patch; an extended streak of milky light; a large resolvable nebula; a very compressed cluster of minute stars hardly discernible; or as an immense collection of large scattered stars of various sizes. And either of these appearances will take place with them according as their own situation is more or less remote from ours.' (Herschel 1786, p. 466.)

However, in 1791 his belief in island universes was shaken and even shattered. His 40-foot reflector had shown him 'a most singular phenomenon! A star of about the 8th magnitude, with a faint luminous atmosphere, of a circular form, and of about 3′ in diameter. The star is perfectly in the center, …' It would later be known as NGC 1514. In Herschel's opinion, 'we … either have a central body which is not a star, or have a star which is involved in a shining fluid, of a nature totally unknown to us.' He opts for the second solution of this riddle. As he has seen more objects of a similar nature, he draws the crucial conclusion: 'Perhaps it has been too hastily surmised that all milky nebulosity, of which there is so much in the heavens, is owing to starlight only.' (Herschel 1791, pp. 82–84.) He then re-examined Orion and found that this nebula was after all a true nebulosity. In 1811, he confessed that his opinion on the construction of the heavens had undergone a gradual change, and he added: 'We may also have surmised nebulae to be no other than clusters of stars disguised by their very great distance, but a longer experience and better acquaintance with the nature of nebulae will not allow a general admission of such a principle, although undoubtedly a cluster of stars may assume a nebulous appearance when it is too remote for us to discern the stars of which it is composed.' (Herschel 1811, p. 270.)

In the following years he put the emphasis very much on the connection between stars and nebulae, and on nebulae out of which stars are born. He

Fig. 4.2. The planetary nebula NGC 1514. With NGC 1514, Herschel realised that not all nebulae could be stellar agglomerations. He found more such 'planetary nebulae' and became convinced that they are stars in formation. (Photograph courtesy of Mike Mandall/NOAO/AURA/NSF.)

obviously was fascinated by the possibility of witnessing the birth of stars. Thus, in 1814, he speculated about 'double nebulae joined by nebulosity between them, and that we have now before us 19 similar objects, with no other difference than that instead of nebulae we have stars with nebulosity remaining between them, should we not surmise that possibly these stars had formerly been highly condensed nebulae, … and were now by gradually increasing condensation turned into small stars; and may not the nebulosity still remaining shew their nebulous origin?' (Herschel 1814, p. 250.)

Whereas in 1784 he cited M31, the great galaxy in Andromeda, as a nebula that shows indications of consisting entirely of stars, in 1814 he grouped it together with M1, the Crab nebula, into a class of objects that 'must at present remain ambiguous'. In addition, he foresaw the breaking up of the Milky Way. He was convinced that stars have been and still are formed by condensation of nebular matter due to gravitation. He thought the Milky Way was in a process of evolution from a former state of widely diffused nebular matter of immense dimension to an irregular agglomeration of globular clusters where the nebular matter had been condensed into a large number of stars. And he added, that although we did not know the rate of going of this mysterious chronometer, it was nevertheless certain that, since the breaking up of the parts of the Milky Way proved that it could not last for ever, it equally bore witness that its past duration could not be infinite (Herschel 1814).

Herschel believed himself to be a witness of evolution, where stars are born and galaxies are formed and dissolved. His early fascination with the island universes gave way to an infatuation with the birth of stars. He was discovering evolution within the Milky Way, and it seems that he lost interest in the construction of the Heavens beyond our own system.

William Herschel died in 1822, but his son John continued the observational tradition. In 1864, John Herschel (1792–1871) published the *General Catalogue of Nebulae and Clusters of Stars*. This catalogue was later extended in 1888 by J. L. E. Dreyer (1852–1926) to become the *New General Catalogue* (NGC). His work on double stars and nebulae catapulted John Herschel to a preeminent scientific status, independent of his being the son of a famous father. Nebulae continued to attract attention, but their nature was still an enigma.

The Leviathan of William Parsons, the Earl of Rosse

In the middle of the nineteenth century, William Parsons, 3rd Earl of Rosse (1800–1867), built his huge telescope in Ireland – the Leviathan as it was unofficially called. The investigation of nebulae was its main purpose. With an aperture of 6 feet and a focal length of 53 feet it remained the biggest telescope in the world until 1917 when the 100-inch Hooker on Mount Wilson began operating (1 foot = 12 inches = 30.5 cm).

In 1850, William Parsons presented to the Royal Society his nebular drawings obtained with his giant telescope. He had discovered a new subclass: some nebulae were of spiral shape. By 1850 he had already found 14 of that kind. But he muses that 'as observations have accumulated the subject has become, to my mind at least, more mysterious and more inapproachable.' (Rosse 1850.) Indeed, what is the qualitative difference between the Orion nebula or a planetary nebula and a spiral nebula?

It had been realised that stars did not all need to have the same absolute brightness, so their apparent brightness ceased to be regarded as a safe indicator

Fig. 4.3. The Leviathan of Parsonstown. The Newtonian telescope of the Earl of Rosse, built in the early 1840s, has a 6-foot aperture and a focal length of 53 feet. The story of the construction is told by Rosse (1861). (Plate XXIV from Rosse 1861.)

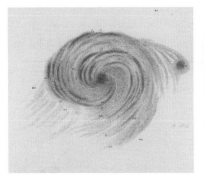

Fig. 4.4. The discovery of spiral nebulae. Left: M51 (NGC 5194), An example of the remarkable quality of observations made with the Leviathan. (Rosse 1850.) Right: M51 photographed by the Hubble Space Telescope.

of relative distances. This discouraged Rosse from offering explanations about the nature of the nebulae. Rosse's most important achievement with his giant telescope was the discovery of spiral structure in certain nebulae. It was these spiral nebulae for which Slipher, from 1912 onwards, eventually found large cosmological redshifts.

Huggins applies spectroscopy on nebulae

In 1864, William Huggins (1824–1910) added an entirely new dimension to nebular research. Fraunhofer in 1814 had observed and published a solar spectrum of very high quality (Fraunhofer 1814/15), and he followed this up with the spectra of some bright stars. Wollaston had already discovered dark lines in the solar spectrum, but he saw them as division lines between different colours. He found 'that, by employing a very narrow pencil of light, 4 primary divisions of the prismatic spectrum may be seen …', and, being aware of the infrared and ultraviolet already discovered, he thought that 'we may distinguish, upon the whole, six species of rays into which the sun beam is divisible by refraction.' (Wollaston 1802.) Fraunhofer's spectrum brought a new quality to spectroscopy.

It was soon realised that stellar spectra are qualitatively like the solar spectrum; they also consist of a continuum, punctured by innumerable absorption

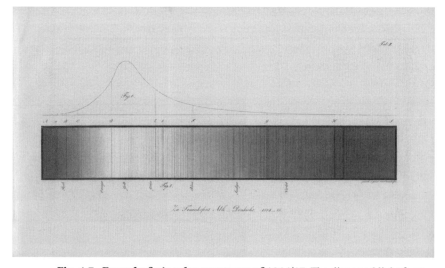

Fig. 4.5. Fraunhofer's solar spectrum of 1814/15. The dispersed light forms a continuum, intercepted by numerous absorption lines. The principal features were designated by Fraunhofer with letters from A to I. The individual absorption lines belong to well-defined elements, and their patterns are now known from the laboratory. The intensity of the solar continuum radiation, disregarding the absorption lines, has been estimated by Fraunhofer in the curve above the spectrum. Fraunhofer marked the colours from 'roth' on the left to 'violet' on the right. On the right, we see two strong dark lines labelled H. Later, they were named H and K; they are absorption lines due to singly ionised calcium at 3934 and 3968 Å (angstrom). They will appear very prominently as H and K lines in the spectra of galaxies (see Chapter 12). The spectrum was hand-drawn by Fraunhofer. (Fraunhofer 1814/15.)

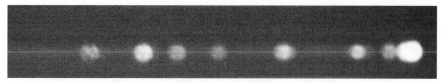

Fig. 4.6. Spectrum of the planetary nebula NGC 6818. Strong emission lines from the nebula, with a very weak continuum from the star. The emission lines are due to hydrogen (H), helium (He), oxygen (O) and neon (Ne). From right to left: [O III] 5007 Å blended with [O III] 4959 Å, Hβ 4861 Å, He II 4686 Å, Hγ 4340 Å and [O III] 4363 Å, Hδ 4101 Å, [Ne III] 3967 Å, [Ne III] 3869 Å and Hε 3970 Å, [O II] 3737 Å. This is a prismatic spectrum, and therefore the dispersion is non-linear. (Lawrence Aller and Lick Observatory, with thanks to Jim Kaler.)

lines. Huggins had become strongly interested in spectroscopy after he had heard about Kirchhoff's epoch-making discovery of 1859. Kirchhoff, in collaboration with Bunsen, had found a fundamental explanation for the relation between gaseous absorption and emission spectra, and they determined the identity of several chemical elements responsible for absorption lines in the solar and stellar spectra.

In 1864, Huggins and Miller reported their observations of stellar spectra in the *Philosophical Transactions*, and Huggins appended a supplement 'On the Spectra of Some of the Nebulae' (Huggins 1864). There were some fundamental issues at stake. They had found a basic similarity among the spectra of fixed stars, and between them and the solar spectrum. Could this similarity be extended to the nebulae? After all, if they really are vast agglomerations of stars, they should show stellar-like spectra. Huggins first focused on the most enigmatic ones, those which William Herschel had called 'planetary nebulae'. No spectrum was seen, just a short line of light perpendicular to the direction of dispersion. His surprise was such that at first he suspected some fault in his instrument. Then, after closer scrutiny, he found two more lines and, eventually, an exceedingly weak continuum.

The spectrum was dominated by what we now know to be a 'forbidden' emission line of twice-ionised oxygen, [O III], at the wavelength of 5007 Å (angstrom), and two additional emission lines, one – Hβ at 4861 Å – being due to hydrogen, the other at 4959 Å, which is again a forbidden line of twice-ionised oxygen. More than sixty years would elapse before Ira Bowen finally identified the enigmatic 'forbidden' lines, designated as [X], in 1927. Whilst the 'normal' lines correspond to electric dipole radiation, the 'forbidden' lines are due to magnetic-dipole or electric-quadrupole transitions. The probabilities of such transitions are exceedingly small and, because in those days they were unobservable in the laboratory, they were called 'forbidden'.

Looking at the planetary nebulae through the telescope and comparing nebular and stellar spectra, it became clear to Huggins that they could not be very distant agglomerations of stars. The emission lines must be due to matter in a gaseous form. This gas might be connected to the central object, which was occasionally seen and most probably responsible for the feeble continuum in the spectrum. Thus, in the middle of the nineteenth century, spectroscopy seemed to support Herschel's hypothesis of the origin of stars and planets as being due to condensation from fluid material.

Huggins had also observed M31, the nebula in Andromeda. 'The brightest part of the great nebula in Andromeda was brought upon the slit. The spectrum could be traced from D to F. The light appeared to cease very abruptly in the orange; this may be due to the smaller luminosity of this part of the spectrum. No indication of the bright lines.' (Huggins 1864, p. 441.) Huggins did not comment further on M31 – the show had been stolen by the planetary nebulae.

In the *Philosophical Transactions* of 1866, Huggins presented a more extended investigation on nebular spectra, comparing them with the appearances of nebulae as described by John Herschel and Rosse. Then he made the distinction between nebulae that show just a few strong emission lines in their spectrum, indicating a gaseous nature, and nebulae whose spectra are apparently continuous (Huggins 1866a).

In August 1866, Huggins delivered a discourse at Nottingham before the British Association for the Advancement of Science. It was immediately translated into French by Abbé F. Moigno and, with a short preface by Huggins, published in Paris (Huggins 1866b). In the same year it was also translated into German and published in Leipzig. It contains his collected nebular experience up to 1866. Huggins recalled that for the previous 150 years astronomers had tried to find the true nature of the nebulae. The interest in them had grown significantly since William Herschel had suggested that they might be part of the original matter, which had served in the formation of stars, and that in some of the nebulae we may still recognise the different phases through which suns and planets passed in their evolution from luminous clouds to their present state.

Huggins' principal conclusions were that there are two classes: those producing an emission-line spectrum and those with a continuous spectrum. Some of the emission-line spectra may contain a very feeble continuum. These nebulae are of a true gaseous nature; they cannot be resolved into individual stars, but they might contain a single star at their centre. Orion and those classified as planetary nebulae belong here. All the nebulae that can be resolved into individual stars in the telescope have a continuous spectrum. But also the nebula in Andromeda belongs here. Its spectrum has many similarities with the great cluster in Hercules. Could it then be that all those nebulae that have a continuous spectrum are stellar agglomerations? A close comparison with the observations of Rosse pointed in that direction. However, in 1866, Huggins refrained from further speculations about the nebulae. The views about the Universe were changing rapidly, he said, and it was wiser to wait for new facts and keep an open mind.

On the spectrum of the great nebula in Andromeda

Towards the end of the nineteenth century, nebulae were out of fashion, but the torch was not extinguished. With a mirror of 32-cm aperture and 96-cm focal length, Scheiner, who worked at the Astrophysical Institute of Potsdam, photographed the spectrum of the Andromeda nebula with an exposure of seven and a half hours. 'The continuous spectrum can be clearly recognized on it from F to H, and faint traces extend far into the ultra-violet.' He found surprising agreement with the solar spectrum, taken with the same apparatus. His conclusions were clear: 'The Andromeda nebula belongs to the class of the spiral nebulae which all give a continuous spectrum. Since the previous suspicion that the spiral nebulae are star clusters is now raised to a certainty, the thought suggests itself of comparing these systems with our stellar system, with special reference to its great similarity to the Andromeda nebula.' He gave additional arguments to show that these spiral nebulae were probably of the same nature as the Milky Way (Scheiner 1899). Scheiner's original photographic plate still exists, but the quality has deteriorated and it does not lend itself to a satisfactory reproduction. However, a scan has been published by Oleak (1995).

Thus, at the end of the eighteenth century, the hypothesis of island universes enjoyed a brief spell of favour when William Herschel thought he had found sound astronomical evidence in its favour. The discovery of planetary nebulae, as true nebular objects with just one central star, stole the limelight. Although never completely abandoned, island universes went slowly out of fashion during the nineteenth century, and nebulae captured interest as possible precursors of stars and planetary systems. Indeed, belief in 'island universes' had dwindled to the extent that in 1890 Agnes Clerke could write: 'No competent thinker, with the whole of the available evidence before him, can now, it is safe to say, maintain any single nebula to be a star system of co-ordinate rank with the Milky Way. A practical certainty has been attained that the entire contents, stellar and nebular, of the sphere belong to one mighty aggregation, ... so far as our capacities of knowledge extend.' (Clerke 1890, p. 368.) The resurrection of the concept of island universes began in 1899 with Scheiner's two-page publication and its authentication by observations.

5

Island universes turn into astronomical facts

A universe of galaxies

Spectroscopy had separated nebulae into two classes, but atomic physics was not yet developed sufficiently enough to help. Nebular spectra had to wait until the late 1920s for a more advanced theoretical interpretation. Photography, however, did much to revive interest in nebulae. In a lecture on spiral nebulae to the Société des Amis de l'Université on 1 February 1912, P. Puiseux – well known for a photographic atlas of the Moon – gave the reason: any prolonged photographic exposure of a spiral nebula leads to the discovery of a great number of similar objects. Whereas up to the year 1900 they were considered a rare species, it now looked as if they formed the majority among the nebulae. It therefore became an urgent matter to find out more about their true nature. Puiseux himself was of the opinion that the great spiral nebulae can be compared to our Milky Way (Puiseux 1912).

We will not list all the individual publications that dealt with nebulae or the size of the Milky Way, but we can obtain an impression of the opinions that were floating around in the scientific community by looking at the views of Curtis, Shapley and Slipher. Curtis and Shapley will be the opponents in the 'Great Debate' of 1920 about the scale of the Universe, and Vesto Slipher will figure prominently in the discovery of the expanding universe.

Distances were the big unknowns: how big is the Milky Way, what is the distance to the spiral nebulae? (Distances will be given either in light years, parsec (pc), or kpc. $1 \text{ kpc} = 1000 \text{ pc}$, $1 \text{ pc} = 3.26$ light years, 1 light year $= 9.5 \times 10^{12}$ km. We usually retain the units employed by the authors cited.) Since Herschel's 1785 publication it had been accepted that the Sun lies close to the centre of the Milky Way. This was still the view of Kapteyn in 1920, although in 1918 Shapley had separated the Sun from the galactic centre by 20 000 parsecs.

Shapley located the nebulae in the outer confines of the Milky Way, but Curtis and Slipher saw them as truly extragalactic systems. The debate could have been terminated in 1922 when Öpik calculated the distance to Andromeda at 450 kpc, which placed it far outside Kapteyn's or Shapley's Milky Way. But no notice was taken. Only in 1925 was the question of island universes finally settled when Hubble placed three nebulae definitely outside the Milky Way, among them Andromeda.

We may also keep in mind that, in 1917, when the debate about island universes accelerated and entered its final phase, Einstein published his cosmo-logical model – see the next chapter – which predicted a finite universe, the size of which could even be determined from existing observations. However, the issues discussed in this chapter were rarely influenced by theoretical cosmology.

Kapteyn's Galaxy

At the beginning of the twentieth century, the Dutch astronomer Jacobus Cornelius Kapteyn (1851–1922) initiated a worldwide programme for investigating the structure of the Milky Way by measuring the positions of a vast number of stars. The result basically confirmed the picture of a disk-shaped stellar agglomeration, with the Sun located close to the centre (Kapteyn and van Rhijn 1920). Thus, in 1920, a large number of astronomers still saw the Milky Way basically as Herschel had sketched it in 1785.

Although the existence of interstellar matter was already acknowledged, its absorbing properties were not known, and the dimming effect on the light from remote stars was hardly taken into account. This omission led to gross under-estimates of the number of stars in the direction of the galactic centre and placed the Sun close to the middle of the Milky Way.

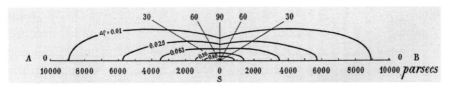

Fig. 5.1 Kapteyn's Galaxy. Distribution of density in a plane perpendicular to the galactic circle. AB is the plane of the galaxy; the numbers 0, 30, 60, 90 are galactic latitudes. The curves are lines of equal stellar density, the density at the Sun (at the centre of the system) being taken as unity. This figure, completed by the addition of its reflected image on the other side of the line AB, represents a complete section through the galactic system at right angles to the galactic plane. (Figure 2 from Kapteyn and Van Rhijn 1920.)

Shapley's changing view of the Milky Way

Harlow Shapley (1885–1972) was a colourful personality: just read his autobiography *Through Rugged Ways to the Stars* (1969). He was one of the most influential American astronomers in the first half of the twentieth century. Shapley started his career in Princeton where, in 1913, he completed his thesis with H. N. Russell on eclipsing binary stars. In 1914 he moved to Mount Wilson, and there he published an impressive series of papers on our galactic system, mainly based on observations with the 60-inch reflector. From 1921 to 1952 he was director of the Harvard College Observatory.

We shall present Shapley's contribution to the disentanglement of the galactic structure in a time-reversed order. We first look at the galactic model that established his reputation. Then we show how he reached his conclusions.

Shapley's Supergalaxy

A picture, completely different from Kapteyn's Galaxy, was constructed and presented by Harlow Shapley. It resulted primarily, but not exclusively, from his study of globular clusters. They are beautiful spherical aggregations with diameters of about one to several hundred light years, containing hundreds of thousands of stars, or well over a million for the largest ones, held together as a cluster by their combined gravitation. There can be little doubt, Shapley said, that the galactic plane, defined by the faint stars and by the Milky Way clouds, is also a symmetrical and fundamental plane for the system of globular clusters. In other words, the distribution of clusters shows that, notwithstanding their great dimensions, they are subordinate members of the far greater galactic system (Shapley 1918a). He concluded that all known celestial objects – stars, nebulae and clusters – are members of a single unit: the Galaxy. It has a diameter of at least 300 000 light years. Its centre lies in the direction of the constellation of Sagittarius, at a distance of about 20 000 pc (65 000 light years) from the Sun. The Sun itself is located about 20 pc north of the galactic plane. He then summarised his work in an article called 'Globular Clusters and the Structure of the Galactic System' (Shapley 1918b). He had employed the Cepheids as indicators for the distances of globular clusters.

In those years the period–luminosity relation of Cepheids had become a crucial tool for determining distances. Classical Cepheids are evolved massive stars in their giant phase. Thus the prototype, δ Cephei, is about 40 times as big as the Sun, has about 5 times its mass, and is about 2000 times as luminous, with a surface temperature varying from 5500 to 6800 K. They vary in brightness very regularly, with periods between 2 and 100 days. The period of δ Cephei is 5.4 days.

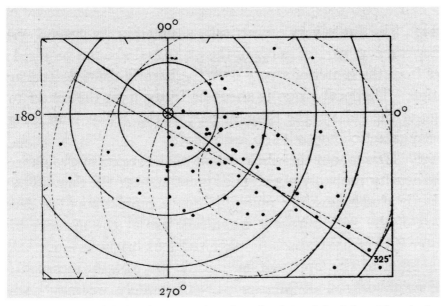

Fig. 5.2 Globular clusters projected on the plane of the Galaxy. The galactic
longitude is indicated for every 30 degrees. The 'local system' is completely within the
smallest full-line circle, which has a radius of 1000 parsecs. The larger full-line circles,
which are also heliocentric, have radii increasing by intervals of 10 000 parsecs. The
broken line indicates the suggested major axis of the system, and the broken circles
are concentric about the galactic centre. Dots indicate the clusters. (Shapley 1918b.)

In 1912, when searching for variables in the Large Magellanic Cloud (LMC),
Henrietta Leavitt (1868–1921) discovered the period–luminosity relation for
Cepheids. She found that their absolute brightness is proportional to the loga-
rithm of the pulsation period. Cepheids of the same pulsation period therefore
have the same absolute luminosity. Comparing their apparent (measured)
luminosities leads immediately to the ratio of their distances.

Much later – in the early 1950s – Walter Baade (1893–1960) found that there are
two classes of Cepheids. The classical Cepheids found by Leavitt in the LMC
are stars with several times the mass of the Sun. The Cepheids found by Shapley
in globular clusters originate from old stars with masses of only half the solar
mass; they are now called population II Cepheids, or W Virginis stars. At equal
period, they are less luminous than the classical Cepheids. The difference between
the two classes did not much influence the island-universe debate, as it was offset
by other mistakes, but the difference was later responsible for underestimating
distances to galaxies and the very high values obtained for the Hubble constant.

Shapley also discussed nebular distances derived from novae. If spirals
were of the same size as the Milky Way – for which he assumed a diameter of
300 000 light years – then their measured angular diameters would place them

at distances greater than a hundred million light years. Novae observed in spirals would, at such enormous distances, have to be intrinsically much brighter than any nova seen in the Milky Way. But if one assumed that those discovered in spirals were of the same character as our own, then their distances placed them just at the outer fringes of our Milky Way.

Shapley's relocation of the Solar System from the centre to the outer confines of the Milky Way had a big impact in and beyond the astronomical community. In 1921, *Die Naturwissenschaften*, a weekly journal on the progress of natural sciences, medicine, and technology, carried a review on Shapley's discoveries by August Kopff, who from 1912 to 1950 was an important figure in German astronomy. This article shows how a European colleague viewed Shapley's world. Kopff recalled that before Shapley's research, the generally accepted picture had been that the totality of observable stars was contained in a flattened, rotationally symmetrical ellipsoidal volume, with the Sun situated fairly close to the centre. This picture had been somewhat amended by Kapteyn, who showed that there was no sharp outer boundary and stellar density was diminishing continuously, faster towards the poles of the system than in the plane.

Shapley's results contradicted that traditional view in several ways. The Milky Way, Kopff says, with its stars and diffuse nebular matter, fills the galactic disk to an extent of 200 000 light years; the extent perpendicular to the plane is 130 000 light years. Our Sun lies in the galactic plane, but at a distance of about 60 000 light years from its centre. The galactic plane is not only the plane of symmetry for the stars in general, but for the stellar clusters as well. Stellar clusters now assume a cosmological significance. They create the framework in a self-contained material island, within which the rest of the stars and the diffuse nebular masses integrate. Only the place of the spiral nebulae was unclear: perhaps they belong to the Milky Way, or they may be systems of the same kind as the Milky Way.

Kopff then linked the observational discoveries to the cosmological models of Einstein and de Sitter (see the next chapter), and added that even with Shapley's size for the Milky Way there was ample space for many more similar systems. But Kopff cautioned: Shapley's results are based on the presumption that everywhere in the Universe the physical constitution of the celestial objects are the same (Kopff 1921).

Shapley thus became associated with a new concept of the Milky Way – the Sun is shifted from the centre to the edge of the Galaxy, and the totality of the observed universe is contained in that Galaxy. This is indeed the picture he advocated in 1918. But, beware, the starting and endpoints of his journey were far apart; it began in 1915.

In 1915, Shapley embarked on a research programme on globular clusters at the Mount Wilson Observatory, based on an extensive investigation of the magnitudes and colours of their individual stars. The second of his long series of communications dealt with M13, the Hercules cluster (Shapley 1916). The observing history of M13 dates back at least to Halley and it was because it had been studied so well that Shapley was attracted to it.

In 1913, von Zeipel had published an extensive theoretical analysis of globular clusters. He considered them to be like gaseous spheres, where the individual stars are the molecules. He assumed a Maxwell–Boltzmann velocity distribution and that most of the stars were bound in binary systems. From the few stellar radial and transverse velocities already known, and from the distribution of the stars that could be counted in the resolved outer regions of the clusters, von Zeipel concluded that globular clusters were gigantic systems, containing approximately a million stars. From the observed transverse proper motion and the spectral radial velocities it was also possible to calculate approximate distances. For M92, another globular cluster in Hercules, he thus obtained a distance of approximately 1000 pc (von Zeipel 1913); we now know it to be about 8000 pc (26 000 light years). Shapley accepted these conclusions: globular clusters are very massive and very distant objects.

Shapley then set out to determine the distance to M13 himself. He saw a high degree of comparability between our galactic system and globular clusters, and therefore assumed that the brightest stars in M13 had the same absolute magnitude as the most luminous stars in our galactic system. Today we know that his assumption was not justified; it led to an overestimate of the distance, which he found to be about 100 000 light years. His estimate of a maximum radius of 10 000 light years for the Milky Way led him to conclude that M13 was far outside the galactic system. He did not give sources for his assumed size of the Galaxy, but it was a number mentioned in the publications of those days. Shapley thought M13 to be about the same size as the Large Magellanic Cloud, so he considered other clusters to be comparable to the Milky Way in size and form. In a footnote, he added: 'The asymmetric arrangement of globular clusters has led Bohlin (1909) to suppose that they form a system at the center of the galactic system and that the Sun is eccentrically situated. The relative distance of clusters and galactic stars precludes the possibility of such an hypothesis.' (Shapley 1916, p. 86.) Let us look at Bohlin's model; as his publication is not easy to access, we summarise its essentials here.

Bohlin had studied the distribution of globular clusters from the catalogues of William and John Herschel, and Dreyer's catalogue of nebulae and clusters.

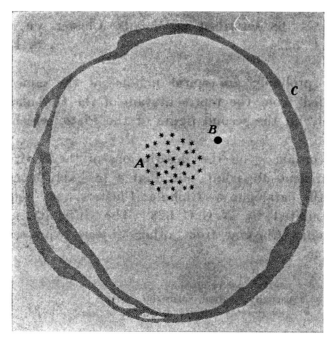

Fig. 5.3 General arrangement of the Galactic System according to Bohlin.
Bohlin's description of the model says: 'A. Globular Clusters' system; B. Sun with the
planetary system; C. Milky Way, including the general system of fixed stars, of which
B is the only one represented in the figure.' (Bohlin 1909.)

His first conclusion was: 'The Globular Clusters System is really situated in the
Centre of the Galactic System.' He continued: 'In fact, if this is supposed, one
needs only to place the sun with the planetary system somewhat beyond the
boundary of the Globular Clusters System in a direction opposite to the apparent place of the centre of the Clusters on the celestial sphere in order to obtain
the aspect actually exhibited by the distribution of the Globular Clusters on the
heavens.' Bohlin then discussed the distribution of stars in the Milky Way, citing
among others, Herschel, Seeliger, de Sitter and Kapteyn, and adopted Kapteyn's
conclusion that the stars show a symmetrical distribution in respect to 'the
Galactic circle'. This 'galactic circle' is prominent in his model (see Fig. 5.3).

In Bohlin's view, the Galaxy was originally a great spheroidal shell of matter,
which then somehow collapsed. He drew parallels to planetary nebulae, which
he thought to be stars forming through condensation of matter out of a luminous shell. Because of their rotation they would condensate into basically
disk-like shapes. 'The Galactic System is supposed to have been such a planetary
nebula, which in the manner above described has broken down at its poles,
being at present at a pretty advanced stage of a Ring Nebula. The centre of

the System is pointed out by a Nucleus consisting of the assemblage of the Globular Clusters. The Spiral Nebulae around its poles are to be considered as the products of the broken parts of the shell, while the Milky Way again is derived from its equatorial belt. Therein exist at present certain irregular gaseous nebulae of great extent, due to latter collisions and frictions, by which the original state of matter, once characteristic of the ring nebula, have been revived.' (Bohlin 1909, p. 20.)

Shapley disagreed; his final summary was: 'There seems to be reasonably clear evidence that this [M13] and similar globular clusters are very distant systems, distinct from our galaxy and perhaps not greatly unlike it in size and form. The open clusters, on the other hand, seem to be relatively small parts of the local system; but further observation is necessary to make certain of this conclusion. (Shapley 1916.) Thus, in 1915, Shapley rejected a Galaxy with the Sun placed off centre. But the position of the Sun was not really the issue of his paper. His main point was that globular clusters did certainly not occupy the centre of the Milky Way. In his view they were not even part of our Galaxy; if anything, they might be galaxies in their own right. Thus he tended to the hypothesis of 'island universes', a hypothesis he was summoned to refute in the 'Great Debate' of 1920.

Although Bohlin's writings were somehow cryptic, and Shapley rightly rejected his suggestions of a Galaxy with globular clusters forming its centre, Bohlin may still be regarded as a precursor to Shapley's model of spherically arranged globular clusters and an off-centre position of the Sun.

Why did Shapley change his mind? After the first two investigations on globular clusters (Mount Wilson Communications 115 and 116) he continued with great vigour to study their colour indices, distances and distribution. When browsing through his numerous publications from 1916 to 1919, the impression grows that two developments were responsible for his conversion: (1) he saw that the globular clusters were arranged symmetrically to the galactic plane, and (2) he realised that the Galaxy was larger than he had originally assumed. Let us briefly comment on both points.

In 1918, Shapley found new ways of determining distances to globular clusters: (1) he calibrated a period–luminosity relation of what he thought to be globular cluster Cepheids. Actually they were RR Lyrae stars, which are of lower luminosity than the Cepheids. Thus, when mistaken for Cepheids, fallaciously they mimic a greater distance. This influenced Shapley's absolute scale but not the pattern of the distribution; (2) with the help of RR Lyrae stars he calibrated the brightest stars in globular clusters, which could then serve for clusters where RR Lyrae stars were too faint; (3) assuming that all globular clusters have the same diameter, the measured apparent diameters could be used to calculate the distances.

Fig. 5.4 M13: The globular cluster in Hercules. M13, one of the brightest globular clusters in the northern sky, lies at a distance of approximately 7700 pc (25 100 light years). In its centre it reaches a density of about 100 stars per pc^3, compared with about 1 in the solar neighbourhood. (Photograph by Noel Carboni, Digitised Sky Survey.)

By 1918, Shapley had measured 32 globular clusters north of the galactic plane and 37 south of it (Shapley 1918b). This large number of clusters, whose distances had been determined in a consistent way, allowed him to find their spherical distribution in space. He located the approximate centre of their distribution in the direction of Sagittarius and at a distance of about 20 000 parsecs.

During those years Shapley's conception of the Galaxy became more elaborate. He also looked at galactic clusters. His fifth paper on stellar clusters was dedicated to M11, one of the richest open clusters. In his conclusion he wrote: 'The magnitudes of the blue stars seem to indicate the remoteness of the star-clouds and also their great dimensions. Similar results have been obtained in a number of other galactic fields, all pointing to a diameter of the galactic system considerably greater than generally supposed.' (Shapley 1917.)

Shapley's view of the Milky Way was changing and the picture of the Galaxy needed revision. Slipher's redshifts of globular clusters showed that seven out of eight of those with high galactic attitudes were approaching the equatorial segment. But Shapley does not find any in the galactic plane – he refused to see the effects of absorption – which suggests to him that upon approaching the galactic regions, globular clusters may be disrupted and transformed into open galactic groups (Shapley 1918a). This provides a connection between the realm of globular clusters and the Milky Way. In the same publication, he enlarged his vision of the Milky Way: 'The stars of spectral type B, according to Charlier, form a flattened system of some 2000 light years radius, which he identifies with the general stellar universe; the group is very small, however, as compared with the system now outlined by globular clusters, and we may assume instead that these B stars comprise a localized stellar organization.' (Shapley 1918a, p. 228.)

In April 1918, Shapley wrote about views expressed on the relation of globular clusters to the galactic organisation: 'Until last year or so most students of stellar problems believed rather vaguely that the sun was not far from the center of the universe, and that the radius of the galactic system was of the order of 1000 parsecs.' (Shapley 1918c.) Shapley had been one of these students.

Thus, with a larger Galaxy, and the globular clusters distributed symmetrically to the Milky Way and even, according to his belief, changing into open clusters when approaching the galactic plane, his Supergalaxy emerged, around which even the spiral nebulae found their places. In 1919, Shapley and his wife published a study where, among other subjects, they accommodated the spiral nebulae into their picture of the world (Shapley and Shapley 1919). Having arrived at the conclusion 'that the galactic system absorbs globular clusters', they contrasted this with nebulae, where 'the brighter spiral nebulae as a class, apparently regardless of the gravitational attraction of the galactic system, are receding from the sun and from the galactic plane.' Concerning their velocities they said: 'The speed of spiral nebulae is dependent to some extent upon apparent brightness, indicating a relation of speed to distance or, possibly, mass.' This was not an anticipation of Hubble's observationally found linear velocity–redshift relationship. In their paper, the Shapleys proposed as a preliminary hypothesis:

> that the discoidal galactic system originated from the combination of spheroidal star clusters and has long been growing into its present enormous size at their expense. The evidence further suggests that the

galactic system now moves as a whole through space, driving the spiral
nebulae before it and absorbing and disintegrating isolated stellar
groups. Apparently the suggested interpretation requires that two types
of sidereal organization prevail generally throughout extra-galactic
space: spiral nebulae, and stars of known types assembled for the most
part into globular clusters; and while globular clusters now known are,
at least potentially, members of the galactic system, the spirals are not
members, rather being general inhabitants of extra-galactic space. The
hypothesis demands that gravitation be the ruling power of stars and
star clusters, and that a repulsive force, radiation pressure or an
equivalent, predominate in the resultant behavior of spiral nebulae.

Shapley's 1915 picture of huge globular clusters as independent galaxies had an
undeniable touch of grandeur. But no less grand was his new picture of an
enormous Galaxy, ploughing through and engulfing everything that was within
the observer's reach, including spiral nebulae. That the Sun was but a tiny spot,
far away from the centre of the newly emerging galactic structure, simply added
to the magnificence of Shapley's universe. Shapley's autobiography leaves no
doubt about who had unsealed this prodigiousness for us to admire: 'I stayed
with the Cepheids and clusters during those early years at Mount Wilson, until
I crashed through on the distances and outlined the structure of the universe.'
(Shapley 1969, p. 52.) Well, when crashing through he overshot by at least a
factor of four, and his universe turned out to be just one galaxy among a myriad
of others. But that does not detract from his merits.

Sandage has looked in detail at Shapley's work on globular clusters and the
structure of the Galaxy, and he pinpoints the many dead ends that can be involved
in such pioneering research (Sandage 2004, Chapter 17). In his Chapter 14 he
also gives a description of the monumental efforts that were needed to reveal
the structure of the Galaxy, where Shapley's efforts represented just one brick in
the building, albeit an important one. In addition, he points out that even in
1922 it seemed quite plausible from the work of Kapteyn and van Rhijn that the
Sun should be close to the centre of the Milky Way.

Slipher favours island universes

In 1917, Slipher published an article with the brief title 'Nebulae' in the
Proceedings of the American Philosophical Society (Slipher 1917). His introduction
shows how uncertain the community was about their nature: 'In addition to
the planets and comets of our solar system and the countless stars of our stellar
system there appear on the sky many cloud-like masses – the nebulae. These for

a long time have been generally regarded as presenting an early stage in the evolution of the stars and of our solar system, and they have been carefully studied and something like 10 000 of them catalogued.'

Doppler shifts and redshifts are central to the cosmological debate, and Slipher was the first to have them systematically observed in nebulae. Light is an electromagnetic wave running through space. The wavelength of a particular colour or special feature in the spectrum is usually designated by λ. When an object moves away from us we see the wavelengths of all the light waves from that star or galaxy lengthened by a common factor. The shift in a particular wavelength is designated as $\Delta\lambda$. There is a simple relationship between the ratio $\Delta\lambda/\lambda$ and v, the velocity of the object: the Doppler shift. It is given by $\Delta\lambda/\lambda = v/c$, where c is the velocity of light. Thus, a simple measurement of the wavelength shift $\Delta\lambda$ gives the speed with which the object is moving away from us.

When Slipher wrote his article, more than half of the nebulae known were in spiral form. He emphasised that 'the faintness of these spectra has discouraged their investigation until recent years'. He had obtained spectrograms of 25 spiral nebulae. By measuring their wavelength shifts, which he associated with Doppler shifts, and thus with radial velocities, he found 4 of them moving towards us, and 21 moving away. Their high velocities surprised him. Stars in our neighbourhood move with average velocities of about 30 km/s, but the average nebular velocity was closer to 570 km/s. This, he remarked, puts the spiral nebulae in a class of their own. Slipher then determined our motion relative to the spiral nebulae and found that we are moving in the direction of $\alpha = 22^{\mathrm{h}}$, $\delta = 22°$ at 700 km/s. He concluded:

> While the number of nebulae is small and their distribution poor this result may still be considered as indicating that we have some such drift through space. For us to have such motion and the stars not show it means that our whole stellar system moves and carries us with it. It has for a long time been suggested that the spiral nebulae are stellar systems seen at great distances. This is the so-called "island universe" theory, which regards our stellar system and the Milky Way as a great spiral nebula which we see from within. This theory, it seems to me, gains favour in the present observations.

Curtis and his novae

Also in 1917, Heber D. Curtis (1872–1942) published a short note with the title 'Novae in spiral nebulae and the island universe theory'. He mentioned the discovery of novae in spiral nebulae and compared their brightness with those in our own Galaxy. 'If we assume equality of absolute magnitude for galactic and spiral novae, then the latter, being apparently 10 magnitudes fainter, are of the order of 100 times as far away as the former. That is, the spirals containing the novae are far outside our stellar system: and these particular spirals are undoubtedly, judging from their comparatively great angular diameters, the nearer spirals.' (Curtis 1917.) Assuming that novae in spirals reach the same brightness as our own at the maximum of their outburst, they would have to be at a distance of twenty million light years if the average distance of our novae is twenty thousand light years.

Was there a villain in the game?

In 1919, Harlow Shapley had no definite view about the nature of the spirals, but he was convinced that his huge Milky Way contained everything that was then within observational reach; perhaps the spirals are true gaseous objects, but if they are stellar systems, they cannot be far outside the Milky Way. One of his main arguments was their internal motion. Van Maanen had identified many condensations in the spiral nebula M101, and compared their locations on photographic plates taken from 1899 to 1915 (Van Maanen 1916). He found that the nebula was rotating. He also applied his technique to other nebulae.

If one knows the distance to the nebula and its size, then the absolute rotational velocity can be found from van Maanen's observations. Shapley pointed out that if spirals, in which rotation had been measured, were located far outside our Milky Way, the observed angular diameters implied large absolute diameters, and their rotational velocities, derived from van Maanen's measurements, would exceed the speed of light (Shapley 1919).

Doubts were voiced about van Maanen's measurements, but he insisted that they were correct. Hubble eventually lost patience with these claims and re-measured some of the plates van Maanen had inspected; the two opponents both worked at the Mount Wilson Observatory. Hubble found no rotation. He then asked two highly regarded observers for an independent opinion: 'At the writer's request, Dr. Nicholson measured one of the pairs for M 81 and Dr. Baade the two pairs for M 51. The displacements in each case agree with the writer's results within the uncertainties of measurement and do not indicate rotations of the order expected.' (Hubble 1935.) Van Maanen's rejoinder was published immediately, adjacent to Hubble's verdict. He remained adamant, although he agreed that his newest measurements resulted in a much lower rotation (Van Maanen 1935). For more inside information on the Van Maanen–Hubble quarrel, see Sandage (2004).

Van Maanen introduced a similar confusion into solar physics as he did for galaxies. Hale started observations of solar magnetic fields in 1908. He expected the Sun to have a simple dipole field. Strengths of magnetic fields were derived from spectral lines known to give large Zeeman-splittings in the strong magnetic fields of sunspots. Up to 1914, more than 2100 photographic plates were taken. Reduction of the observational data was a delicate matter, and whether a measurable splitting was found depended on the person doing the data reduction. Van Maanen found what Hale had expected and was therefore considered to be the most reliable. From 1930 to 1933 a new series of observations were taken. Van Maanen was no longer a member of the team, and no measurable general magnetic field was detected. This prompted Hale to postulate the general solar magnetic field to be variable with an unknown period.

Because of the magnetic field's fundamental significance for solar physics, Stenflo decided to re-measure the old plates. To avoid a personal bias he made automatic recordings of the plates and analysed the data on a computer (Stenflo 1970, the information in this section is based on that article). Stenflo summarised the main results thus: 'In contrast to the neat behaviour of the field measured by van Maanen, the present remeasurements of the same plates do not show any significant field at all at any of the latitudes.' Stenflo later commented: 'The detailed comparison left me with the impression that wishful thinking must have played a considerable part in van Maanen's data reduction.' (Personal communication, 2007.)

The Great Debate

On 26 April 1920, the American National Academy of Science staged a discussion meeting on 'The Scale of the Universe'. The contest was between the concept of a rather small Milky Way and spiral nebulae as island universes (defended by Curtis), and the concept of a huge Milky Way, containing all astronomical objects known at that time (defended by Shapley). This encounter is now known as the 'Great Debate'. Its outcome was inconclusive. Both sides had good arguments, but often for the wrong reasons. The discussion continued for another four years.

We can now see where the two sides were right or wrong. The Great Debate has been discussed many times and commented upon. We just mention a few main points. Shapley had put his money on the right horse when he employed globular clusters to determine the size and centre of the Milky Way. The Sun is indeed closer to the edge than the centre, and on this point Curtis was wrong. But, for several reasons, Shapley overestimated the distances to his globular clusters, and therefore overestimated the size of the Galaxy by at least a factor of

five. Curtis put his money on another winner when he emphasised the significance of novae for determining distances. He also did the right thing by distin-
guishing between normal novae in Andromeda and the special case of S
Andromeda, which had flared up in 1885; it was a supernova, but the fundamental distinction between novae and supernovae had not yet been made. The
surprisingly fast rotation of spiral galaxies, as apparently seen by van Maanen,
was indeed a crucial argument, and Shapley gave it much weight. Curtis disregarded van Maanen's claims. He thought that such determinations were too
difficult to be done reliably. He was right, but no one has ever found out what
van Maanen did wrong. Transcripts of the addresses by Curtis and Shapley during
the Great Debate can be found in the *Bulletin of the National Research Council*, Vol. 2,
on pages 171–217. For a critical evaluation of that text, see Hoskin (1976).

Öpik finds the distance to Andromeda

In 1922, Öpik calculated a distance to the Andromeda nebula M31. He
based his derivation on the observed rotational motion derived from Doppler
shifts and on the virial theorem. If v is the velocity along a stable circular orbit
at a distance r from the centre of the Galaxy, there is a gravitational pull due to
the mass inside r, which is compensated by the centripetal acceleration. Öpik
assumed that the energy radiated per unit mass is the same in Andromeda as in
the Milky Way. He calculated the mass from the observed apparent luminosity
and the run of the velocities along the radius r from the measured Doppler
shifts. The radial distance r can be expressed through its observed angular size
and the distance to Andromeda. In this way, Öpik determined the distance to
Andromeda to be 450 000 pc (Öpik 1922).

Öpik's distance is closer to the presently accepted distance of 800 000 pc than
Hubble's distance of 285 000 pc, which he published in 1925. However, Hubble
made the headlines, whereas Öpik did not. Öpik concluded his very clever
derivation with the assessment that his result 'increased the probability that
this nebula is a stellar universe, comparable with our Galaxy.'

Hubble cuts the Gordian knot

The 33rd meeting of the American Astronomical Society was held in
Washington from Tuesday 30 December 1924 to Thursday 1 January 1925. The
proceedings are reported in *Popular Astronomy*, Vol. 33, page 158. Among the
members in attendance we find H. D. Curtis and H. Shapley, but not E. P. Hubble.
In the report a special mention is reserved for an event on Thursday, 1 January
1925. We read:

Fig. 5.5 Hubble and the Mount Wilson reflector telescope. Left: Edwin Hubble (1889–1953). Right: The Hooker 100-inch (2.5 m) telescope on Mount Wilson. It went into operation in 1918 (first light, November 1917). Hubble did his most significant observations with this instrument. The Hooker telescope dominated observational cosmology until the 200-inch Hale reflector on Mount Palomar became operational in 1948/49. (Photographs courtesy of California Institute of Technology; Observatories of the Carnegie Institution, Washington, Mount Wilson Observatory.)

It was in this session that Professor Russell presented the communication by Dr. Edwin P. Hubble on "Cepheids in Spiral Nebulae," which was to share in the joint award of the thousand-dollar prize given for an outstanding paper at the Washington meeting. Dr. Hubble, working with the 100-inch Mount Wilson reflector, had succeeded in resolving portions of two of the spiral nebulae, those of Andromeda and Triangulum, into separate stars, and from a study of the period-luminosity curves of Cepheid variables in the nebulae had derived distances approaching one million light years for each, thus bringing confirmation to the so-called island universe theory.

Hubble's report and findings were printed in the proceedings and in several other publications (e.g. Hubble 1925a,b,c). An illuminating account of some of the circumstances leading to Hubble's announcement is given in Berendzen and Hoskin (1971). That account also recalls that, even in 1925, van Maanen's rotating spirals were still looming over the nebular debate.

Hubble had been appointed to Mount Wilson in 1919. With the 60-inch and the 100-inch Hooker he had the world's most powerful telescopes at his disposal, and he made good use of them. He looked for and found some novae in

Andromeda, but even more important was his discovery of Cepheids. They provided him with the key to the distances of extragalactic nebulae. We have just heard about his Washington announcement on 1 January 1925. By making a comparison with other distance indicators, Hubble had seen that the period–luminosity relation worked well for the 11 Cepheids he had found in NGC 6822, which he described as a faint cluster of stars and nebulae (Hubble 1925a). This allowed him to assign NGC 6822 to a region definitely outside our galactic system. He then published his observations of 22 Cepheids in M33, and 12 in M31 (Hubble 1925b,c). Hubble had adapted Shapley's period–luminosity curve for the Cepheids to photographic magnitudes and found a distance of 285 kpc for the two spiral nebulae. This put them far outside the Milky Way, even when accepting Shapley's exaggerated size of the Galaxy.

Those publications of 1925 definitively settled and closed the debate on island universes. It was now generally accepted that spiral nebulae are galaxies, similar to our Milky Way. The vast number of spirals detected in the otherwise seemingly empty immenseness of space qualified them as building blocks of the visible universe.

6

The early cosmology of Einstein and de Sitter

In parallel with the gradual emergence of convincing evidence for a universe composed of galaxies, theoretical cosmology began in 1917 with Einstein's static universe as a grand entrance (Einstein 1917). Einstein aimed at nothing less than unveiling the structure of the Universe. Combining his crowning achievement, General Relativity, first published in 1915 (Einstein 1915a,b, 1916), with the assumption that matter was at rest and evenly distributed throughout the Universe, the Cosmos became a curved, finite space. And, what to some must have looked like blasphemy, Einstein even gave the formulae to calculate the size and total mass of the Universe. In the same year, just a few months later, de Sitter published another, even more provocative relativistic model (de Sitter 1917). His universe contained no matter, and therefore seemed to be remote from reality, but astronomers became highly intrigued. De Sitter predicted redshifted spectra for cosmologically distant objects, and thus gave the scientific community a prospective explanation for Slipher's observed redshifts. However, the theory was embarrassingly enigmatic. From 1917 to 1931, the year when the expanding universe finally became generally accepted, a considerable number of theoretical investigations were published, in particular by Einstein, de Sitter, Eddington, Friedmann, Lanczos, Weyl, Lemaître, Robertson and Tolman, and most of them dealt with de Sitter's model.

All the models were derived from the fundamental equations of General Relativity, the Einstein field equations (it is customary to set the speed of light c to $c = 1$):

$$R_{ij} - \frac{1}{2}g_{ij}R = 8\pi G T_{ij}.$$

They are often written as

$$G_{ij} = 8\pi G T_{ij}.$$

Qualitatively, the left-hand side contains the description of the geometrical structure of *spacetime*, and the right-hand side with the gravitational constant G and the energy–momentum tensor, T_{ij}, contains information on the material content of the Universe. Thus they relate the structure of space – in the sense of the 4-dimensional spacetime – to its matter content, matter being understood to include anything to which mass or energy is attributed. These equations are explained in more detail in the Mathematical Appendix.

In this chapter we look only at the models of Einstein and de Sitter, and we discuss some of the early controversies. Moreover, we give a heuristic explanation of some fundamental relations. However, in order to place these initial efforts within a historical flow, we briefly outline the subsequent developments; they will be further itemised in later chapters.

The early theoretical work of Einstein and de Sitter was built on the obvious hypothesis of an unchanging, static universe. But, in 1922, Alexander Friedmann, who then worked in Petrograd in the Soviet Union, showed that the most natural solution of Einstein's field equations resulted in a dynamical universe. No one paid attention, except Einstein, who dismissed Friedmann's results, first as faulty, then as irrelevant. However, observational evidence for a dynamic universe was already on stage, albeit waiting in the wings. Since 1912, Slipher had been obtaining reliable redshifts from the spectra of spiral nebulae, but the distances to the nebulae were a sore point. Thus, the efforts of Wirtz, Silberstein, Lundmark and Strömberg to unravel de Sitter's predictions of certain distance–redshift relations did not succeed. A breakthrough became feasible with Hubble's systematic distance investigations in his voluminous publication of 1926 (Hubble 1926).

In 1927, after a first attempt in 1925, Lemaître had a fresh look at the cosmological models of Einstein and de Sitter. He found a crucial flaw in de Sitter's construct. Combining theory and observations, he showed that we live in an expanding universe. His article was published in French and in a journal that was not read widely. When, in the autumn of 1927, Lemaître showed it to Einstein, the response was as negative as it had been towards Friedmann.

In England, the astrophysical group around Eddington was very active in discussing cosmological concepts. Eddington was one of the foremost proponents of the general theory of relativity. In 1919, he had been instrumental in organising the solar eclipse observations that verified Einstein's prediction of how curved space would bend a beam of light that passed close to the Sun. The success of that expedition established Einstein's fame and made him a public hero. Eddington was an astronomer and physicist, a true astrophysicist. His *The Mathematical Theory of Relativity* was first published in 1923. The second edition followed in 1924. It was reprinted many times and became a standard textbook on relativity. It included Slipher's redshifts, which thereby entered the cosmological mainstream.

Late in 1929, Eddington and de Sitter were again in close contact. Hubble had observationally found a linear relation between nebular redshifts and their distances, and de Sitter had confirmed the news by doing his own study. In January 1930, Eddington and de Sitter discussed the consequences, but found no satisfactory interpretation. Approximately two months later they learnt about Lemaître's expanding universe, and about his prediction of a linear velocity–distance relationship. From that moment on, things moved fast: the hypothesis of the expanding universe was gaining general acceptance.

Some fundamental relations
(Also see the Mathematical Appendix)

We present here some concepts that will appear in the discussion on the theoretical aspects of cosmology. This is outside the narrative, which continues with the next section on 'The static universe of Einstein'. The reader may choose to skip this brief tour on key concepts and return to it when needed.

In September 1932, Sir Arthur Eddington delivered a public lecture at the meeting of the International Astronomical Union at Cambridge (Massachusetts). There he said: 'When a physicist refers to curvature of space he at once falls under suspicion of talking metaphysics.' (Eddington 1933, Chapter II.) He then tried to make curved space palatable to the general public. But he realised that his brave attempts might not attain the desired goal and admits: 'In fact, the pictorial conception of space-curvature falls between two stools: it is too abstruse to convey much illumination to the non-mathematician, whilst the mathematician practically ignores it and relies on the more dependable and more powerful algebraic methods of investigating this property of physical space.' – Well, don't ever give up.

General Relativity

General Relativity is the unified theory of space, time and gravitation; it links physics to geometry. It describes the Universe as a 4-dimensional space-time construct, whose geometrical properties, such as curvature, are related to the distribution of mass and energy. General Relativity extends Newton's theory of gravitation as well as Newton's equations of motion.

Although General Relativity is a mathematically difficult subject, a major obstacle against really assimilating this theory is the concept of curved spacetime.

We are used to thinking of the Earth, the stars, and galaxies as 3-dimensional objects existing in a boundless 3-dimensional space, which acts mentally as a 3-dimensional 'container' for all these 3-dimensional objects. Einstein does away with that container. There is no rigid absolute space against which we can define motion and remains the same, whether we purge it of all matter or fill it to the brink. Space – the 3-dimensional spatial part as well as the 4-dimensional spacetime – is structured by the distribution of matter; at the same time the structure of space influences the distribution of matter. Thus, the structure of space and the distribution of matter are intrinsically related. When constructing cosmological models, the structure of space is derived from the distribution and state of matter using Einsteins's field equations.

Note that the term 'space' is used differently in different connotations. We talk about the mathematical n-dimensional space, of which the 4-dimensional spacetime of General Relativity is one example. 'Space' may also be used for the 3-dimensional spatial world. The meaning should always be clear from the context.

Curved spacetime

In cosmology, time and the 3-dimensional spatial world are united into a 4-dimensional spacetime. General solutions of Einstein's field equations do not separate these four dimensions into one time and three spatial dimensions. However, astronomical observations suggest a homogeneous and isotropic universe. In this case, the 4-dimensional spacetime splits naturally into a 3-dimensional spatial and a 1-dimensional time fraction, similar to that of Newtonian physics. We are free to single out a particular instance in time, and to look at the structure of the 3-dimensional spatial fraction. In the static universe of Einstein that structure is the same for all times, but in an expanding universe it changes continuously. So, let us for the moment concentrate on the 3-dimensional spatial fraction.

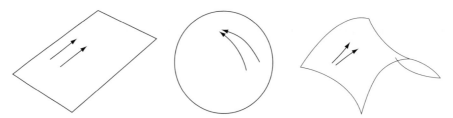

Fig. 6.1 Flat and curved space. Left: Flat space, where Euclidean geometry is valid. Centre: Positively curved space, where spherical geometry applies. Right: Negatively curved space, where hyperbolic geometry is valid.

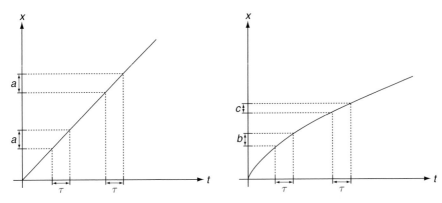

Fig. 6.2 Curved time. The coordinates give time t and distance x. Left: A static universe, no curvature in time. During a constant time interval, τ, light always covers the same fraction, a, of the distance between two galaxies. Right: Curved time of an expanding universe. During the same time interval τ, light covers different fractions, b and c, of the distance between two galaxies; the fraction b is larger than the fraction c of the later time.

When in the Newtonian world we want to describe the path that light has taken between two points A and B, we can employ Euclidean geometry to find it. However, in relativistic cosmology, space is structured differently. Depending on the matter–energy content, we may deal with flat, positively curved, or negatively curved spaces.

Before discussing cosmological models, let us look at a 2-dimensional example of curved space. Among many differently curved and shaped geometrical spaces, the following three types play an important role in cosmology: flat, spherical and hyperbolic. We direct our attention to distances: What is the shortest distance between two points A and B? Such a path is called a geodesic. (Geodesics are a fundamental concept to which we will return.) Figure 6.1 shows that the shortest connection between two points is a straight line for the flat space, a section of a great circle on the sphere, and a hyperbolic curve in the negatively curved space. We also see that in the flat and the hyperbolic cases the geodesic is a fraction of an infinitely long curve: the straight line and the hyperbola. In the spherical case it is a fraction of a great circle of finite length, within a finite space. In this case, the line returns to its point of departure, but it arrives there from the opposite direction of departure; this is an example where a geodesic can be the shortest or the longest distance between two points. We also find that in flat and negatively curved spaces, parallel lines never cross, whereas they do intersect in positively curved space.

When the theory of General Relativity is applied to cosmology, it turns out that our 3-dimensional spatial world can be structured either according to flat, spherical or hyperbolic geometry.

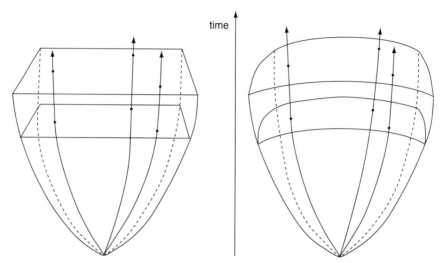

Fig. 6.3 Projection from three into two spatial dimensions. The Universe evolves in time from a near singularity. The world-lines of three spatially motionless galaxies are given. The increase in separation is due to the expansion of space as time moves forward. Left: Schematic presentation of a spatially flat universe, curved in time. It corresponds to the Einstein–de Sitter universe of Chapter 14. Right: Section of an expanding spatially curved and closed universe, as proposed by Lemaître in 1927.

We now return to the general case of curved spacetime. The structure of our Universe could be such that the 4-dimensional spacetime is curved, however, the spatial fraction has a flat geometry. The Einstein–de Sitter universe that we shall meet later is such a case. We may then talk of a universe where space is flat but time is curved. Let us look at the significance of 'flat' or 'curved' time. In a static universe, space always remains the same, and a beam of light always needs the same time for covering the distance between two given objects that remain at rest. However, if that universe expands, the time needed for covering the distance between the same two galaxies increases. This is shown in Fig. 6.2. It also has to be remembered that during expansion the length of the path between two galaxies is continuously changing.

Let us now proceed to different cases of 4-dimensional spacetimes. How can we draw a 4-dimensional picture of spacetime on a piece of paper? We cannot, because a sheet of paper has only two dimensions. However, when a Renaissance painter wanted to convey the impression of a 3-dimensional world on his 2-dimensional canvas, he employed perspective drawing; when an architect hands to the builder his plans for a new house, they consist of projections from a 3-dimensional space on to a 2-dimensional plan. To visualise cosmological

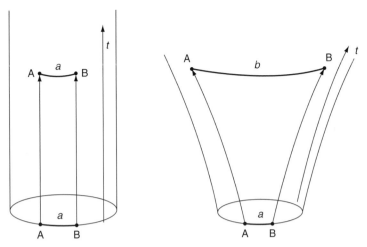

Fig. 6.4 Projection from three into one spatial dimension. The arrows indicate the motion in time within the spacetime of two observers A and B who remain spatially immobile as time passes. The spatially curved and closed 3-dimensional fraction of spacetime is represented by a 1-dimensional circle. In the static universe, the distance between the observers A and B remains constant, whereas it increases in the expanding universe. Be aware that only the mantle of the cylinder and the surface of the hyperboloid represent the corresponding universes. Thus, any motion in time and space can only occur on the mantle of the cylinder or the surface of the hyperboloid. The model on the left corresponds to Einstein's original model of 1917, the one on the right to Lemaître's expanding closed universe of 1927.

models we can use similar methods. How then could we represent spacetime on a piece of paper? With perspective, we can draw 3 dimensions on to the paper. We have to reserve one dimension for time, so 2 dimensions are left for the spatial world.

As examples, we show in Fig. 6.3 two different homogeneous universes, both expanding in time. The 3-dimensional space can be projected on to a 2-dimensional plane. The evolution in time of the spatial distances can be shown with a succession of slices, corresponding to successive states of 'cosmic time' (the temperature of the cosmic background radiation might serve as a universal clock). The first one is an open, spatially flat universe. Euclidean geometry is valid for the 3-dimensional spatial part, but not for the 4-dimensional spacetime. Only two spatial dimensions are shown in the picture. The second example is spatially curved and closed. Its projection on to a 2-dimensional plane results in a homogeneous sphere. We show the evolution in time of a fraction of the sphere. With the evolution in time, the spatial curvature of the curved universe changes.

We now push the projections one step further and project the 3-dimensional spatial world into one dimension (Fig. 6.4). Instead of the 2-dimensional surfaces, a single line will represent the spatial fraction of the Universe. For the closed universes of Einstein's static model and Lemaître's expanding model, the Universe is now represented as a circle. For the static universe, this circle always remains the same when time moves forward. For Lemaître's expanding universe, the circle is growing in diameter: the history of the Universe – the 4-dimensional spacetime – is now represented as a hyperboloid.

Events, world-lines, geodesics and proper time

In General Relativity we talk about events. An event is a point in spacetime. It can mark the presence of an object or a point in a given place at a given time. In 4-dimensional spacetime, an object always moves in time, even if it is motionless in space: the observer then just moves along the time direction. In the absence of any other influence except gravitation, such a motion in time is uniquely determined by the geometry of the model. The path of an object in spacetime is called a world-line.

Geodesics are an important concept in cosmology. A geodesic is the shortest or longest connection between two points in spacetime; it is a generalisation of the straight line. Light always moves on geodesics. A geodesic is the path along which a body moves in 4-dimensional spacetime when only gravitation and no other forces act on it. In Euclidean space, a geodesic is a straight line. In curved space, geodesics are curved lines. For a given cosmological model, geodesics are calculated from an equation involving the metric: the geodesic equation.

A world-line can be a geodesic. Let us look at an example. When descending from the Eiffel tower I move in space and in time: in spacetime. If I take the stairs, I then move on a world-line which is not a geodesic, because gravity is permanently counteracted by the supporting power of the stairs. If I jump in free fall from the top of the Eiffel tower, I also move on a world-line; however, this time the world-line is a geodesic.

Time can be defined in many ways. The most important definition is proper time. Proper time is what an atomic clock shows to an observer moving along a world-line.

Cosmological models

Cosmological models involve a geometrical concept and assumptions about the physics of the material content. The 4-dimensional spacetime is a geometrical space. Its structure is given by the metric, which is a solution of Einstein's equations. At present, it is not known how to solve these equations in full generality. In order to describe physical reality, basic assumptions about the structure of spacetime have to be made; they reduce the complexity of full generality and identify specific solutions. Such assumptions are usually suggested by observations or philosophical considerations. An important assumption is homogeneity. It is based on a combination of observations and some outrageous extrapolations; it is the generalisation of the Copernican principle, also known as the 'cosmological principle', which says: there is nothing special about our place in the Universe.

Einstein's field equations

The field equations are a set of equations that determine the structure and evolution of spacetime. They link these properties to the matter and energy content. Without entering into mathematical subtleties, we can define the line element ds as the distance between two infinitesimally close points in spacetime; the size of ds does not depend on the coordinates. The metric is the expression that allows us to calculate the length ds in a given coordinate system. The metric is a tensor. Its components are denoted by g_{ij}; for details, see the Mathematical Appendix.

Static, stationary and dynamic world models

At the time of our pioneers, the mathematics of General Relativity had not been as far developed as it is now, nor were certain physical notions unambiguously defined, such as the concepts of stationary and static spacetimes.

A static spacetime is a special case of a stationary spacetime. It is the simplest version: nothing happens. In the static universe, galaxies remain where they are and distances between them remain fixed for all eternity; the metric tensors, g_{ij}, are the same everywhere and remain the same for all times. Einstein's model of 1917 was of that kind.

For a stationary universe, it is possible to choose a coordinate system in which all components of the metric tensor, g_{ij}, are independent of time. A well-known astrophysical example of a stationary state is the simplified model of dust formation in the steady wind of a red giant: the star loses mass in a steady, radially outflowing wind. At a fixed distance from the surface, dust is formed in the wind and is carried away with it. Thus, physical conditions may be different at different locations, but they do not change in time. In cosmology, the Steady State model of Hoyle, Bondi and Gold represents a stationary universe: new matter is continuously created, but expansion of the Universe keeps the density constant. The metric tensor, g_{ij}, remains the same for all locations and for all times. Yet, due to the expansion, the distances between existing galaxies increase.

Another example: an assembly of non-rotating stars at rest is a static system. It corresponds to a certain structure of spacetime. The same assembly, but with one of the stars rotating steadily, represents a stationary state; the corresponding structure of spacetime is different from the non-rotating case.

When Einstein and de Sitter published their cosmological concepts in 1917, they intended to describe a static world. The dynamical spacetime of the models of Friedmann and Lemaître is neither static nor stationary. The large-scale physical properties change as time advances, which result in a change of the metric with time. In popular explanations, the closed, expanding universe is represented as an expanding balloon, where the total amount of matter is preserved. For a given moment, physical conditions are the same everywhere. With time running forward, density and curvature decrease and, accordingly, the metric changes as well.

When discussing the original papers we shall employ the terminology of the respective authors. We have to keep in mind that the terms static and stationary, as defined above, were employed more liberally. However, it should always be clear from the context what was meant in these early publications.

Coordinate systems

Coordinate systems are introduced for the actual description of positions and motions in spacetime. The choice of a particular coordinate system is arbitrary. Some systems are mathematically more convenient; others are more helpful for visualisation. This freedom of choice may lead to confusion, but one should keep in mind that only those statements that are independent of the arbitrarily chosen coordinate system can be considered as physically relevant.

The static universe of Einstein

Basic properties

Relativistic cosmology began with Einstein's publication of 1917, where he theoretically investigated the large-scale structure of our Universe. He set out from the simplest assumption that still seemed compatible with observations: the Universe is static; it remains in time as it has always been. The only force that acts in such a universe is gravitation, which leads to a collapse. To preserve stability, Einstein introduced the cosmological term, now generally called Λ. From the field equations of his general theory of relativity he found a solution that represented a static universe. It is a finite, spherical universe. Its intrinsic structure is most naturally described by non-Euclidean spherical geometry. In Einstein's static, spherical universe there is a simple relation between the mean material density, the cosmological term, Λ, and the radius of curvature, R, which could be called the 'radius of the Universe'. The formulae are given in the Mathematical Appendix.

The Machian principle

In his endeavour, Einstein was much driven by what he later called the Machian principle: the total matter of the Universe is responsible for inertia. This is shown by a publication of 1918, where he explicitly singled out the Machian principle as one among the three main principles of General Relativity.

Einstein lists the three principles (Einstein 1918c), they are (translated from German):

(a) The principle of relativity: the laws of nature are only statements about spacetime coincidences; therefore their only natural expressions are generally covariant equations.

(b) The principle of equivalence: inertia and gravitation are essentially the same [wesensgleich]. From this, and from the results of special relativity, it follows necessarily that the symmetric "fundamental tensor" ($g_{\mu\nu}$) determines the metric properties of space, the behaviour of inertia of the bodies in it, as well as the effects of gravitation. The state of space described by the fundamental tensor we designate as "G-field".

(c) The Machian principle: the G-field is entirely determined by the mass of the bodies. As, according to special relativity, mass and energy are the same and energy is formally described by the symmetric energy tensor ($T_{\mu\nu}$), this means that the G-field is due to and determined [bedingt und bestimmt] by the energy tensor of matter.

Although Einstein would later dissociate himself from the Machian principle, it was his guide in those years.

In 1917, he insisted that in a consistent theory of relativity there can be no inertia relative to 'space', but only inertia of the masses against each other, hence: 'If I remove a mass sufficiently far from all the other masses of the world, its inertia must drop to zero.' (Einstein 1917, translated from German.) Such conditions should be mirrored in the metric tensor, g_{ij}. If there is only a single particle in the Universe, then, according to Einstein's reasoning of 1917, that particle is devoid of inertia. This last point was soon to become a bone of contention with de Sitter.

Einstein's way to his cosmology

Einstein described the tortuous way to his cosmological model (Einstein 1917). He assumed the Universe to be static and evenly populated with stars, resulting in a mean density of matter. 'Der metrische Charakter (Krümmung)

des vierdimensionalen raumzeitlichen Kontinuums wird nach der allgemeinen
Relativitätstheorie in jedem Punkte durch die daselbst befindliche Materie und
deren Zustand bestimmt.' (According to General Relativity, the metric character
(curvature) of the 4-dimensional spacetime continuum is in each point specified
by the matter and its state at that point.) The mean density therefore gives him
the metric tensor, with the geometrical structure of the Universe and its spatial
curvature. But a differential equation on its own does not yet specify a solution:
boundary conditions have to be provided as well.

Einstein starts his cosmological investigation with arguments from classical
physics. If the picture of a Maxwell–Boltzmann distribution for a gas is applied
to the Universe, and stars are considered as the atoms of that gas, then a steady
Newtonian stellar system cannot exist: the finite potential difference between
the centre and the spatially infinite corresponds to a finite ratio of densities. But
if the density of matter vanishes at infinity, then it will also vanish in the centre.
There was also a deadly problem. If an infinite universe is filled with a homogeneous distribution of stars, then the slightest inhomogeneity will lead to an
initially local and eventually universal collapse. To counteract such a cata-
strophe, Einstein introduced his famous cosmological constant. It counteracts
gravitation and plays a crucial part in determining the geometry of the Universe.

Einstein prepares us for the introduction of the cosmological constant:
'Es geht aus dem bisher Gesagten hervor, daß mir das Aufstellen von
Grenzbedingungen für das räumlich Unendliche nicht gelungen ist … Wenn
es … möglich wäre, die Welt als ein nach seinen räumlichen Erstreckungen
geschlossenes Kontinuum anzusehen, dann hätte man überhaupt keine solchen
Grenzbedingungen nötig.' (From what has been said up to now, it is clear, that
I did not succeed in setting up boundary conditions for the spatially infinite.
If, however, it were possible, to see the world as a spatially closed continuum,
then one would not need such boundary conditions.) Indeed, the cosmological
constant, Λ, not only counteracts gravitation, but it also gives a constant curvature to the Universe, which becomes spatially closed. Einstein's universe of 1917
is finite in extent, yet it has no boundary. He did not start with the assumption of
a closed universe, but it dawned on him that a closed universe was a way out of
his boundary problems.

At the beginning of this chapter we mentioned that in Einstein's field equation, $G_{ij} = 8\pi G T_{ij}$, the left-hand side describes the geometry of spacetime, whereas
the right-hand side describes the material and energy content. Einstein added
the cosmological constant, Λ, on the left. One might just as well add it on the
right. Mathematically the placement is irrelevant, but it lends itself to different
interpretations. The cosmological constant may be seen either as a geometrical
property of spacetime, or as an additional energy.

Einstein's universe is homogeneous in space as well as in time: the Universe is the same at any time. We can therefore separate the 4-dimensional spacetime into the 3-dimensional spatial fraction, which will remain the same for all times, and the 1-dimensional time. For the spatial part, the same geometry is valid – as on a sphere. The magnitude of the spatial curvature is determined by the amount of homogeneously distributed mass and energy.

If we wish to represent the 4-dimensional spacetime of Einstein's model, we have to reintroduce time. To fit this onto a 2-dimensional paper, where one dimension is reserved for time, we project the 3-dimensional spatial fraction onto a 1-dimensional circle. The essential information is the radius of the circle, which corresponds to the spatial curvature of Einstein's universe. This circle glides along the axis of time and thus forms the mantle of a cylinder. This is known as Einstein's cylindrical universe: a 4-dimensional spacetime, where the 3-dimensional spatial fraction, now represented as a circle, moves along the axis of time.

In a circle, the radius allows us to calculate the circumference of the circle, and for a sphere the radius allows to calculate the area of the surface. In the same way, knowledge of the radius of curvature, R, allows us to calculate the size of the 3-dimensional spatial fraction of the Universe and its volume. And, if we know R and the mean material density, that is the number or mass of the particles per cm^3, we can calculate the total number of particles or the total mass of the Universe. As mentioned before, there exists a simple relation between the cosmological constant, Λ, the radius of curvature, R, the mean density of the Universe and its total mass. The equilibrium of the static universe is maintained by the balance between the gravitational contraction due to matter and the counteracting cosmological term, Λ.

With the formulae of Einstein's model we only need to know the mean density of matter – which can be obtained observationally – in order to calculate the size of the Universe, its total amount of matter and the cosmological constant, Λ. Thus, that model provides fantastic answers to age-old questions about the nature of our Universe. This universe is homogeneous and isotropic, it has been the same for all the times past and it will remain the same for all future times. Time itself is running at the same rate at every point; this will be different in de Sitter's model.

Eddington was full of admiration for Einstein's daring invention: 'Einstein made a slight amendment to his law to meet certain difficulties that he encountered in his theory. There was just one place where the theory did not seem to work properly, and this was – infinity. I think Einstein showed his greatness in the simple and drastic way in which he disposed of difficulties at infinity. He abolished infinity. He slightly altered his equation so as to make space at great distances bend round until it closed up.' (Eddington 1933, p. 21.)

The newly discovered possibility of calculating global properties of the Universe was fascinating. In Section 69 of his *The Mathematical Theory of Relativity*, Eddington looked at Einstein's formulae connecting the cosmological constant, the observable mean density, the radius of curvature and the total mass of the Universe. He argued that the radius of curvature could hardly be less than 10^{18} km ($\approx$32 000 pc) because the distances of some of the globular clusters are higher. From this he derived a total mass of the Universe of at least 10^{18} solar masses (Eddington 1924). This argument shows how badly the content, as well as the structure and size of the Universe, were known at the time.

Hubble gave an observationally better-founded guess in 1926. His observations had extended the known Universe to much larger distances. He found a mean density of 1.5×10^{-31} grams per cubic centimetre. Using this density, he obtained a radius of 2.7×10^{10} parsecs and a total mass of 9×10^{22} solar masses for the Universe. (We shall come back to his derivation of these numbers.) Hubble commented that the 100-inch telescope could detect normal galaxies to distances of approximately 4.4×10^7 parsecs, or about 1/600 of the radius of the Universe. With better observing equipment 'it may become possible to observe an appreciable fraction of the Einstein universe' (Hubble 1926). It was indeed a fascinating prospect to explore the whole Universe out to its remotest confines.

The static universe of de Sitter

A few months after Einstein's publication, the Dutchman, Willem de Sitter, published another version of an apparently static universe, also built on General Relativity (de Sitter 1917). The most amazing feature was that it is devoid of matter. He retained the cosmological constant and also obtained a world of constant positive curvature. Without matter, nothing can happen, so de Sitter's metric coefficients, g_{ij}, remain constant in time and his universe is static. This paper of 25 pages is not easy to read; in Eddington's words, Einstein's spherical space is commonplace compared with de Sitter's (Eddington 1924, p. 168). However, his model had a great influence on the cosmological debate, which only faded when Lemaître's expanding universe was definitely accepted in the early 1930s.

In the ensuing years there was much discussion about whether or not de Sitter's model really represented a static universe, and this caused much confusion. In 1924, Eddington remarked: 'It is sometimes urged against de Sitter's world that it becomes non-static as soon as any matter is inserted in it. But this property is perhaps rather in favour of de Sitter's theory than against it.' (Eddington 1924.) The confusion was clarified when, in 1927, Lemaître showed that the static world-lines in de Sitter's model are not geodesics (Lemaître 1925a,b, 1927). But up to 1930 that work remained unnoticed.

De Sitter attacked Einstein's adherence to the school of Mach, where ' "inertia" is produced by the whole of the world-matter, and "gravitation" by its local deviations from homogeneity.' To refute such a point of view, he showed that Einstein's fundamental equations have solutions even for density zero, and thus for a universe without matter. Yet, this empty universe had qualities of inertia.

Is de Sitter's universe really static?

Apart from the apparently preposterous quality of containing no matter, de Sitter's original model had another striking property: as soon as matter is put into his universe, it becomes non-static. If an observer and a test particle are placed in this empty universe, the test particle is accelerated and moves away from the observer, without the observer being active in any way. The acceleration is proportional to the product, $\Lambda \cdot r$; it is thus increasing with Λ and with the distance r between the observer and test particle. If the accelerated particle sends out a beam of light, the motion results in a classical Doppler redshift. This effect was not usually much emphasised; if it should be active, it would empty the space around us, and that was incompatible with a static universe and with observations. In a footnote to the German translation of *The Mathematical Theory of Relativity*, Eddington wonders whether at the extremes of our Milky Way this cosmological repulsion term could surpass the gravitational attraction and thus set an upper limit to the size of a stellar assembly (Eddington 1925, p. 237).

There is a second effect: a slowing down of time. In de Sitter's universe, time runs fastest at the location of the observer. With increasing distance from the observer, time runs slower. This implies that light waves emitted at increasing distances will be shifted increasingly to the red, although neither the emitter nor the observer moves. The effect increases quadratically with distance. This de Sitter redshift haunted the cosmological debate from 1917 up to 1930. Could it be that it corresponded to the huge redshifts observed by Slipher in spiral galaxies? Lemaître would solve the riddle in 1927.

De Sitter's Trojan horse

Before we show how the astronomical community reacted to Einstein and de Sitter, we give a few examples of the ensuing discussions among theoreticians.

The models of Einstein and de Sitter were studied and discussed by fellow theoreticians. De Sitter's conceptually difficult model was the main target. Einstein was horrified by de Sitter's solution. He rejected it with the following argument: 'Bestände die De Sittersche Lösung überall zu Recht, so würde damit gezeigt sein, daß der durch die Einführung des "λ-Gliedes" von mir beabsichtigte Zweck nicht erreicht wäre. Nach meiner Meinung bildet die allgemeine Relativitätstheorie nämlich nur dann ein befriedigendes System, wenn nach ihr die physikalischen Qualitäten des Raumes allein durch die Materie vollständig bestimmt werden. Es darf also kein $g_{\mu\nu}$-Feld, d.h. kein Raum-Zeit-Kontinuum, möglich sein ohne Materie, welche es erzeugt.' (Einstein 1918a.) (If de Sitter's solution were valid everywhere, it would show that the purpose which I intended when introducing the 'λ-term' has not been achieved. According to my opinion, the general theory of relativity only then represents a satisfactory system, if it explains the physical properties of space completely through matter alone. There must not be any $g_{\mu\nu}$-field, i.e. no space-time continuum, without matter that produces this field.)

Einstein finished that note with the fundamental objection that, according to his opinion, de Sitter's universe was not empty at all; the matter was all concentrated at the horizon. It will be shown in the chapter on Lemaître, how much of the confusion surrounding de Sitter's model was caused by ill-chosen coordinates, which became singular at the horizon.

In 1918, Weyl, in the very first edition of his *Raum, Zeit, Materie*, and then again in an article of 1919, had already insisted that Einstein was right with his statement that de Sitter's universe cannot be totally empty. Weyl thought that an empty world was in conflict with the laws of Nature, and he further found that Einstein's equations implicitly demanded that de Sitter's empty space was surrounded by a mass-horizon (Weyl 1918 p. 225, 1919). In those years, relativistic cosmology and the interpretation of de Sitter's model were discussed in a lively correspondence particularly between Einstein, de Sitter, Klein and Weyl (CPAE 1998). The language of the discussion is theoretical, and there was no direct reference to astronomical facts. It would have been difficult to find relevant facts, anyway. Except for some large redshifts of spiral nebulae, nothing was available that could have given any clue to the large-scale structure and dynamics of the Universe.

Two years later, Lanczos asserted that Weyl's conclusions were erroneous (Lanczos 1922). Weyl was not amused and replied that he – Weyl – was right (Weyl 1923b). We need not go into details; they were walking on interpretational quicksand. There is a telling footnote in Lanczos' publication (translated from German): 'It is interesting to observe how one and the same geometry can lead to totally different interpretations, depending on the choice of the coordinate system, and how the individual coordinates are interpreted.' This was certainly true for de Sitter's model with its various interpretations. The absence of matter allows many choices of vector fields along which fundamental observers move, and each of these choices suggests a correspondingly different coordinate system.

Lanczos' stationary model

In that same publication, Lanczos wrote down a formal solution of a spatially closed dynamical universe, just as Friedmann had done shortly before and Lemaître would do five years later. However, unlike Friedmann and Lemaître, Lanczos did not grasp the physical significance and he did not consider a non-stationary universe. The concept of a static world was deep-rooted. Yet, Lanczos' critique of de Sitter's model was certainly appreciated by Lemaître, who refers to it in 1927 (Lemaître 1927, footnote 2).

Lanczos did not reply to Weyl's reprimand, but in 1924 published a long article about a stationary cosmology. His first sentences were (translated from German): 'In the sense of general relativity a world is to be considered as stationary when the coefficients of its metric are independent of time, if we are in a coordinate system where the masses on average are stationary. The addition about the coordinate system is essential, because time in itself represents not an invariant concept, unless in its quality as proper time.' (Lanczos 1924.) He wanted his coordinate system to be composed of geodesic lines. He pointed out that de Sitter's 'static' universe was not a stationary world because when a test particle is placed in that model it does not remain at rest. (We shall come back to that aspect when presenting Lemaître's model of 1927.) He added that the field equations without Λ do not have stationary, spherically symmetrical solutions, and that Einstein's solution is the only possible one, when Λ is included. But he disliked Einstein's world because of the tight relation between Λ and the total mass of the Universe, which he considered too much of a coincidence. Lanczos then split the 4-dimensional spacetime into three spatial dimensions, and the time dimension in such a way that his time coordinate corresponded to geodesic world-lines. He arrived at the conclusion that in a stationary metric of de Sitter's universe there is no redshift due to the metric. His own model is stationary, the spherical symmetry has gone, and its mass density has to be higher than a certain minimum, given by the value of Λ in Einstein's model. As Lanczos did not offer any explanation for the observed redshifts, his work did not catch on.

Weyl's contribution

Weyl's contribution shows how theoreticians in the 1920s struggled to disentangle de Sitter's model and create a physically and mathematically coherent concept. For those not used to thinking in their terminology it is often difficult to follow the arguments. However, the reader may take comfort from the fact that the discussion was confusing even to insiders, as testified by remarks of Eddington, Lanczos, Robertson and Tolman. Still, let us have a look.

In 1923, Weyl published a note, 'Zur allgemeinen Relativitätstheorie', where he wanted to find his own way to de Sitter's redshift (Weyl 1923c). Weyl was a mathematician and this shows in his choice of coordinates. When treating de Sitter's hyperboloid, a common mortal would separate time and space by allocating the axis of rotational symmetry to time. Yet, Weyl's choice was different: time was no longer 'our' time, and space no longer 'our' space.

Weyl first recalled that when light travels from a location of a lower gravitational potential to a higher potential, a Doppler shift results. He then studied what happens to light emitted from a star at some point in spacetime and observed by a receiver at another point in spacetime. He pointed out that all the events in the Universe that can be seen by a given observer should be causally connected. Below we shall return to this fundamental physical idea. In order to investigate this 'causally connected world', Weyl chose a set of coordinates that were mathematically well suited to that task, but they did not allow an easy interpretation of redshifts.

Weyl imbedded the 4-dimensional universe with a constant radius of curvature in a 5-dimensional Euclidean space. The Universe is then represented as a 4-dimensional hyperboloid. The geodesic lines are cut out by the 2-dimensional planes through the origin of the 5-dimensional space. As mentioned above, there were no longer any pure time and space coordinates. This made an intuitive comprehension even more difficult. In this representation, Weyl found a redshift, which, for small distances between the light source and the observer, is nearly linearly dependent on the distance.

He also found that 'All stars of our system … flee radially from an arbitrarily chosen observer star; matter has an inherent universal tendency to recede [Fliehtendenz], which finds its expression in the "cosmological term" of Einstein's gravitational law …' (translated from German). From this 'receding velocity' he found another redshift, which, however, differed from the redshift found above (see the Mathematical Appendix). Weyl did not comment on the discrepancy.

Although Weyl did not solve the redshift problem, he brought up the fundamental question of causality. He pondered about the regions of spacetime for which a connection between past and future is possible. In the fifth edition of *Raum, Zeit, Materie*, Weyl described the Universe as a de Sitter hyperboloid. All astronomical objects observed by a given observer belong to a causally connected world (eine einzige zusammenhängende Wirkungswelt). The world-lines of the stars and the observer belong to a single diverging bundle of geodesics. This bundle of geodesics, diverging from a common origin, became known as

Weyl's postulate. Their divergence attests to a universal tendency of matter to disperse due to the action of Λ. Weyl contrasts this to the classical, elementary cosmology in an infinite Euclidean space, and he also sets it against Einstein's cosmology of a closed space with Λ just sufficient to compensate the gravitational contraction, where the world-lines of the stars are characterised by their static coordinates being constant (Weyl 1923a, Anhang III). – The concept of the 'causally connected world' has become important in General Relativity.

Compared with his writings of 1918/1919, Weyl's attitude towards de Sitter's model had changed. 'The principle that matter produces the metric field cannot be maintained in the sense that far from any matter, or if matter has been annihilated, there will be no guiding field, that is the field will be undetermined … The status of rest [Ruh-Zustand] of the metric field is the homogeneous metric, as it reigns on de Sitter's hyperboloid. In agreement with experience I therefore assume, that far from any matter we find this homogeneous state … It will be useful to revive the old picture of the Ether. It will not reappear as a substantial medium, but in the sense that the status of the Ether stands for the existing metric field and electromagnetic field. The relation between matter and Ether is not that of genitor and generated, but rather like a boat on a quiet lake, matter arouses the Ether, which by itself is in a state of rest.' (Weyl 1923a, p. 296, translated from German.) Weyl, however, adds that a final decision about the validity of either Einstein's or de Sitter's model had still to be postponed.

Starting around 1918, Weyl, Einstein, Eddington and others had been dreaming of unifying gravitation and electromagnetism. Einstein had shown how the gravitational field could be incorporated into the structure of spacetime. Could the electromagnetic field not be absorbed in a similar way? In 1921, Eddington had published 'A generalisation of Weyl's theory of the electromagnetic and gravitational fields' (Eddington 1921). For the month of May 1923, we find in the archives of ETH (Swiss Federal Institute of Technology) a brief exchange of letters and postcards between Weyl, who lived in Zürich, and Einstein, who lived in Berlin. They revolve around two subjects: the cosmological term, and the introduction of the electromagnetic field into the field equations.

On two postcards, the first dated 19 May 1923, the second 22 May, Weyl first resolved a problem about the mutual importance of the Maxwell tensor and the cosmological constant already discussed in a letter. In the latter, he talked about a strange effect concerning the changing of sign of the Christoffel symbol when exchanging in the Maxwell energy–momentum tensor positive by negative electricity. In order to avoid that effect, he proposed a purely imaginary cosmological term. (The Christoffel symbol, usually written as Γ, represents the gravitational field and consists of derivatives of the metric tensor, g_{ij}.)

Einstein's frustration

We are not surprised that Einstein began to feel uneasy about the cosmological term and about recklessly playing around with formulae in order to solve physical problems. His answers to Weyl came in two postcards, the first of which was written on 23 May, the second on 26 May. The first message ends: 'Ich sende Ihnen dann die Korrektur meiner Arbeit, die ich unbedingt publizieren muss, weil der Eddington'sche Gedanke notwendig zu Ende gedacht werden muss. Ich glaube jetzt auch, dass alle diese Versuche auf rein formaler Basis die physikalische Erkenntnis nicht weiter bringen werden. Vielleicht hat die Feldtheorie schon alles hergegeben, was in ihren Möglichkeiten liegt. Inbezug auf das kosmologische Problem bin ich nicht Ihrer Meinung. Nach De Sitter laufen zwei genügend voneinander entfernte materielle Punkte beschleunigt auseinander. Wenn schon keine quasi-statische Welt, dann fort mit dem kosmologischen Glied.' (I shall send you the correction of my work, which I absolutely must publish, because Eddington's idea needs to be thought through to the end. I now believe that all these attempts on a purely formal basis will not advance the physical insight. Perhaps the field theory has already given all that it can give. Concerning the cosmological problem I am not of your opinion. According to de Sitter, two material points sufficiently far apart diverge in an accelerated manner. If there is no quasi-static world, then away with the cosmological term.)

The second postcard contains the following passage: 'Im Ganzen ist eine ziemlich resignierte Stimmung bezüglich des ganzen Problems bei mir eingekehrt. Die Mathematik ist schön und gut, aber die Natur führt uns an der Nase herum.' (On the whole I am in a rather resigned mood about the whole problem. Mathematics is all very well, but Nature leads us by the nose.)

Einstein's remarks show, in addition to his preoccupation with a unified theory, utter frustration about de Sitter's theory. Mathematically there was no serious problem. However, the formalism obscured the physical interpretation and he was obviously fed up with playing around with varying sets of coordinate systems. With his 1918 response to de Sitter, he thought he had disposed of that unphysical universe without matter (Einstein 1918a). Now Weyl revived de Sitter. Einstein had introduced the cosmological constant in order to guarantee a static universe, but Weyl holds against him that, although the Universe is empty, test particles will rush away from any observer because of the cosmological constant. We can sympathise with his exclamation: 'If there is no quasi-static world, then away with the cosmological term.' Indeed, if Λ could not fulfil its task, then – according to Einstein – it had lost its right to existence. Eddington and Lemaître would a few years later defend Λ as one of the most fundamental ingredients of modern cosmology.

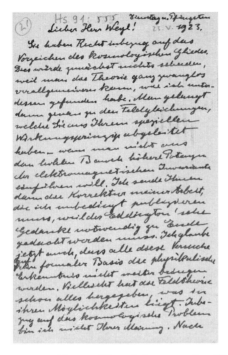

Fig. 6.5 Einstein's postcard to Weyl. Written on Tuesday before Whitsun, which corresponds to 23 May 1923. Note the postage of 180 Mark; this is a witness of the beginning of the proper German hyperinflation. (ETH-Bibliothek, Zurich, Einstein Archiv.)

Einstein was not the only one to feel uncomfortable about de Sitter's Pandora's box. Eddington's comments show that he was not happy with the situation in cosmology, and several years later Robertson expressed the hope to have found a mathematical solution in which many of de Sitter's apparent paradoxes were eliminated (Robertson 1928). It is all the more astonishing that Einstein, in 1922, and again in 1923, brushed aside Friedmann's dynamical solution of the Einstein equations. He did the same in 1927 when Lemaître showed him his discovery of the expanding universe, corroborated with theoretical and observational evidence. We will come back to these two episodes.

The interpretation of the de Sitter type models was never easy, having to do with the concept of time, and at least as much with the freedom to play with coordinate systems. When discussing Einstein's and his models (models A and B respectively in his terminology), de Sitter said: 'In both systems A and B it is always possible, at every point of the four-dimensional space-time, to find systems of reference in which the $g_{\mu\nu}$ depend only on one space-variable (the "radius-vector"), and not on "time". In the system A the "time" of these systems of reference is the same always and everywhere, in B it is not. In B there is no

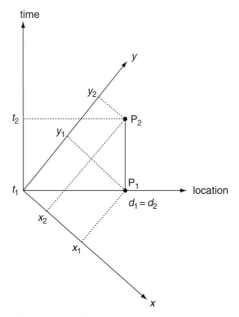

Fig. 6.6 Coordinate systems in transformation. An object in spacetime at the point P_1 with the coordinates (d_1, t_1) remains physically at rest, thus the spatial location d_1 does not change. However, along the time axis the object moves to spacetime P_2 with the coordinates $(d_2 = d_1, t_2)$. In the location–time coordinate system, space and time are neatly separated, and we can easily visualise a spatially motionless motion in time. We can represent the same process in the coordinate system x, y. Mathematically, the motion from the point (x_1, y_1) to the point (x_2, y_2) is equivalent to the move from (d_1, t_1) to (d_2, t_2), and physically it is the same move. However, in the x, y representation the connection to our intuition is lost.

universal time; there is no essential difference between the "time" and the other three co-ordinates. None of them has any real physical meaning. In A, on the other hand, the time is essentially different from the space-variables.' (de Sitter 1917, p. 11.)

These theoretical discussions were only marginally influenced by observations, nor did they much influence observations. There is one exception, de Sitter's initial prediction of two different redshift–distance relations did incite several observers to look for such effects. However, before 1925, astronomical observations were, with the exception of Slipher's redshifts, simply not in a position to contribute any decisive evidence for or against different cosmological models.

Lanczos provides a good example of dreaming about hopeless astronomical tests (Lanczos 1924). To his theoretical investigation he appended some suggestions how his world geometry could be tested by observing the real Universe.

He thought about comparing Euclidean and non-Euclidean models. In a Euclidean universe, the projection of a cosmologically distant sphere will be a circle, whereas in his non-Euclidean universe it will be an ellipse. He suggested that observing the shapes of globular clusters could be of help. He also hoped that stellar statistics could provide a clue to the cosmological structure of the whole Universe. This was written at a time when the astronomical community had not yet reached agreement about 'island universes'. Lanczos was probably under Shapley's spell; his Milky Way incorporated the whole Universe, and he thought that globular clusters were among the most remote objects of the Universe. Yet globular clusters belong, cosmologically speaking, to our immediate neighbourhood and cosmological effects are not practically measurable. His suggestions did not have the slightest chance of testing world models.

No energy conservation in relativistic cosmology!

Contrary to one of the most deep-rooted convictions of classical physics, backed by experiment, no universal conservation law of energy or linear and angular momentum exists in General Relativity. The concepts of energy and momentum can at present not even be defined in a general way. This may appear to be very disturbing. A closer look shows the reasons. In General Relativity, the equivalence principle tells us that the energy of a gravitational field cannot be localised. Einstein's thought experiment of the elevator in free fall is an example.

Gravity keeps us firmly attached to, or pulls us back to the ground when we jump up. Gravity puts weight on us. Is it possible not to feel this gravitational force? Yes, bungee jumping, for example, which is a free fall. But, with our eyes open, we would rationalise that, relative to the cliff, we are accelerating towards the bottom of the canyon. Now we enter Einstein's lift, without windows and in free fall, and we go to sleep in there. When we wake up and do not feel gravity any more, we cannot tell whether during our sleep the lift has been transported so far away from the Earth that no gravity is active, or whether we are in free fall close to the surface of the Earth. According to General Relativity, the inhabitant of the windowless lift cannot distinguish whether he or she is in free fall in a gravitational field, or whether there is no active gravitational field. This is Einstein's equivalence principle. It can be re-phrased as follows: there is no way for an observer to distinguish locally between gravity and acceleration. An observer sitting in a lift that is in free fall feels no gravitational force.

In classical physics, space and time are absolute concepts, and space is totally unrelated to the dynamics of particles within that space. Kinetic and potential energy are well defined relative to that space. However, contrary to classical

physics with its Newtonian absolute space, General Relativity provides no absolute reference system against which we could define kinetic and potential
energy. The spacetime metric, g, of General Relativity describes both the background spacetime structure as well as the dynamical aspects of the gravitational
field. We know of no 'natural way' to decompose it into background and
dynamical parts, where one could then assign energy to the dynamical aspect
of gravitation rather than to the background spacetime structure.

Although on a general level it is not possible to define total energy, nor linear
or angular momentum, we can define them for certain isolated systems. Isolated
gravitating systems, such as binary stars, clusters of stars, galaxies, etc., can be
described in General Relativity by asymptotically flat solutions of the Einstein
equations. They can be thought of as having an asymptotically flat region. For
which type of asymptotically flat isolated systems energy, as well as linear and
angular momentum can be defined, has occupied generations of relativists, and
is still today a subject of research. Thus, in cosmology there is in general no way
of defining energy except for very special cases. Yet the early cosmologists were
mainly led by the idea of energy conservation, and therefore constructed models
where the energy was supposed to be conserved. In some instances it was
justified. An example of a cosmological model, where energy is conserved, is
Einstein's closed and static universe.

Although the early publications on cosmology, such as those of Einstein and
Lemaître, took conservation of energy for granted, it was not clear how energy
should be defined. While Einstein formulated an energy–momentum theorem
for his closed universe, most of his colleagues did not agree with its formulation:
'While the general relativity theory was approved by most theoretical physicists
and mathematicians, almost all colleagues object to my formulation of the
energy–momentum theorem.' (Einstein 1918b, translated from German.)

In more technical terms, the arguments run as follows: his formulation of
the energy–momentum theorem includes quantities that are not tensors, only
pseudo-tensors. This means that they might change 'properties' after certain
transformations. Moreover, his 'energy components' might be transformed
away by choosing an appropriate coordinate system. However, Einstein argued
that these objections were not valid if the Universe was a closed system, which
at that time he favoured. He emphasised that energy and momentum for such a
universe are as well determined by his formulation as in classical physics, and
that they are independent of the coordinates chosen if the motion of the whole
system relative to the coordinate system is given. Thus, for instance, the rest
energy of any closed system is independent of the choice of coordinates.

Weyl cautioned against this attitude in his book, *Raum, Zeit, Materie*.
Introducing variables as energy components of the gravitational field that are

not tensors is, in general, physically meaningless. By changing the coordinate systems, their values can be changed and even be brought to zero. This does not exclude the possibility that such variables can, under special circumstances, represent conserved quantities (Weyl 1921, p. 246). Thus, as mentioned above, the rest energy of any closed system is independent of the choice of coordinates. Einstein's static closed universe represents such a system.

Schwarzschild's vision of curved space

Even before Einstein postulated a spatially closed universe, Karl Schwarzschild (1873–1916) dreamt at the very end of the nineteenth century about such a cosmology. He wondered how the world would look in a non-Euclidean finite space going back onto itself (in sich zurücklaufend). He admitted that such speculations were leading into a geometrical fairyland, but the beauty of it was that one did not know whether or not it corresponded to reality.

Schwarzschild was under the impression that our Milky Way contained all the stars of the Universe. He argued about stellar parallaxes, and came to the conclusion that at present we cannot decide which geometry is valid for our world. But his wish was for a closed universe, because it would be a most comforting feeling if we could assume space to be closed and filled by our stellar system. If that were the case, a time would come when the whole space would have been explored (Schwarzschild 1900). – Schwarzschild died in 1916.

7

The dynamical universe of Friedmann

Whilst Einstein was unhappy about de Sitter's universe without matter, and Lanczos mused about the many ways of interpreting cosmological models, Alexander Friedmann was offering a solution to their worries. However, no one was there to pick up the scent.

Alexander Friedmann was born in St Petersburg in 1888. He spent his most productive years in Petrograd and died in Leningrad in 1925 from a typhoid fever he had contracted on a journey from the Crimea. His father was a ballet dancer in the famous company of the St Petersburg Theatre. His mother seems to have studied the piano at the Conservatory. Alexander's parents married when his father was 19 and his mother was 16. Their marriage was dissolved in 1896 when Alexander was 8 years old. The boy grew up in the family of his father, who had re-married. The young Friedmann studied physics. He was well aware of the progress being made in the world of physics. At university he benefited from a weekly seminar given by Paul Ehrenfest who, in the autumn of 1906, had moved to St Petersburg with his Russian wife, whom he had met in Göttingen. Friedmann maintained contact with Ehrenfest, even after the couple had left for Holland in 1912. The principal topics of the seminars were quantum mechanics, which at that time was basically the atom of Rutherford, Bohr and Sommerfeld, statistical mechanics and relativity. In 1911, Friedmann married, and in 1913 he obtained employment at the Main Physical Observatory in Pavlovsk, where he worked as a meteorologist. After World War I and the transformation of Russia into the Soviet Union he was nominated to a professorship at the Physical Institute in what had become Petrograd. From about 1920 onwards, he collaborated in a regular seminar on General Relativity (Tropp *et al.* 1993).

Fig. 7.1 Alexander Friedmann (1888–1925). Friedmann was the first to publish dynamical cosmological solutions of Einstein's field equations.

An alternative to Einstein and de Sitter

'Über die Krümmung des Raumes' (On the curvature of space) was the title of Friedmann's publication of 1922, where he presented an alternative to the static universe of Einstein and de Sitter. At the outset he accepted Einstein's fundamental equations. He retained the hypothesis of a homogeneous, isotropic 4-dimensional curved universe, and he also kept the cosmological constant, Λ, but qualified it as a supernumerary that could take any value, including zero. The fundamental modification of freeing himself from the imposition of a static universe led to a variable spatial curvature, which at a given time is the same everywhere, but is allowed to vary in time. The parameter of curvature, R, is then no longer a constant, but becomes a time-dependent function $R(t)$. The flow of time is independent of spatial coordinates, thus every point in the Universe has the same proper time. From the field equations he derived a set of solutions, giving an expanding, contracting or oscillating universe, which included as special cases the static model of Einstein, as well as de Sitter's empty universe

(Friedmann 1922). Thus, he had found cosmological solutions representing a dynamical universe.

Friedmann did not get bogged down by the apparently intractable contradictions of de Sitter's model. He began from scratch by (1) assuming the validity of Einstein's fundamental equations, with the cosmological term, and (2) assuming the proper motion of matter relative to each other to be small compared with the velocity of light. He split the 4-dimensional spacetime into space and time, such that at equal 'cosmic time' the Universe is everywhere the same.

Having found a general solution, he first turned to the static cases of Einstein's cylindrical and de Sitter's spherical worlds. He found that they are the only stationary solutions of Einstein's fundamental equations. Then he went on to the non-stationary worlds. In 1922, he limited his study to solutions with positive spatial curvature.

Let us summarise the main new insights that Friedmann published in 1922. If the radius of curvature is allowed to vary in time, there are an infinite number of solutions. Depending on the ratio of Λ against the gravitational effect of the total mass of the Universe, the world can either grow or shrink, or vary in size periodically. If Λ is above a critical value, the Universe can grow steadily from size zero. If Λ is below a critical but still positive value, the Universe grows monotonously, starting from a finite size. If $\Lambda = 0$ there will be an oscillating universe. The oscillation period depends on the total mass of the Universe. We shall discuss some of these solutions in Chapter 17, 'The seed for the Big Bang'.

Friedmann calls the time that has elapsed since the moment when the radius of the Universe, $R(t)$, was zero 'die Zeit seit der Erschaffung der Welt' (the time since the creation of the world). There is, however, no indication that he meant this in a religious sense. The cyclic universe seems to have held a special charm for him. Toward the end of his book, *The World as Space and Time*, he explained the evolution of the Universe, its growth and shrinking, and then said, 'One cannot help thinking of the tales from the Indian mythology with their periods of life. There is also the possibility to talk about the creation of the world out of nothing. But all this has at the moment to be considered as a curiosity which cannot be properly verified by the inadequate astronomical observations.' (Friedmann 1923.) His formulae and thoughts included the Big Bang and the Big Crunch, but he did not develop these possibilities in any detail.

Friedmann did not comment on how one could observationally distinguish between his world models. It was not a lack of interest that deterred him from building a bridge to astronomical observations. In 1922, he commented (translated from German): 'Our knowledge is totally insufficient for doing numerical calculations, and for deciding what kind of world our universe is.' It is futile to speculate how he would have incorporated Slipher's redshifts, had he been

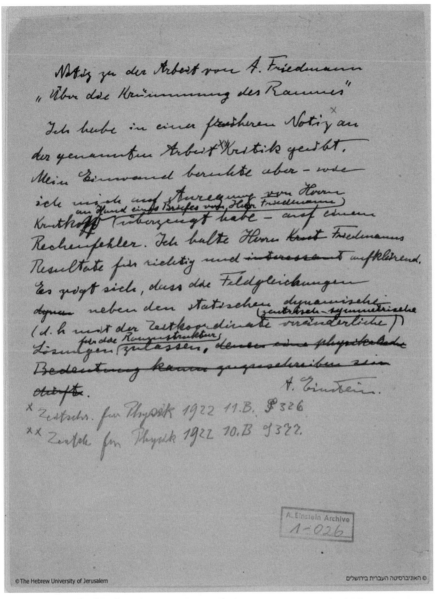

Fig. 7.2 Einstein's retraction. Einstein admits that his criticism of Friedmann's work (Friedmann 1922) was not justified, but it was the fault of a calculation error. This draft was published in *Zeitschrift für Physik* (Einstein 1923a). The cancelled last passage provides a glimpse into Einstein's attitude towards Friedmann's solution. (The Albert Einstein Archives. Einstein 1923b.)

aware of them. He did, however, calculate the age of the Universe from his cyclic model. To find an order of magnitude, Friedmann assumed a universe with $\Lambda = 0$ and a total mass of 5×10^{21} solar masses, without telling us the source for that number, and obtained 10 billion years (10^{10}) for the length of an oscillation period, which he then calls the age of the Universe. But he added: 'These figures can only be considered as an illustration of our calculations.'

In 1924, Friedmann returned to the cosmological scene (Friedmann 1924). This time he looked at world models with negative spatial curvature. Whereas a constant positive curvature leads to a finite universe, a negative curvature gives an infinite universe. He found that for positive mass densities no static world-model with negative curvature exists. Non-stationary worlds are compatible with positive and negative curvature. However, if the curvature is negative there will be no closed, finite world of positive density. Friedmann did not mention world models with zero curvatures. As before, in 1922, Friedmann found no bridge to observations.

Friedmann also pointed to the limits of Einstein's field equations. They do not fully tell us whether the Universe is finite or infinite. The curvature on its own does not settle that question. Additional definitions about the concept of finite and infinite spaces have to be specified (Friedmann 1924).

Einstein's rejection

Friedmann's solutions of the field equations came to Einstein's attention. A few months after their publication, Einstein, in a short note of eleven lines, disqualified them as not compatible with the field equations (Einstein 1922). Eight months later, Einstein retracted his statement in an equally short note, without any further comment (Einstein 1923a).

Considering Einstein's dismissive remarks to Lemaître a few years later (see Chapter 9 on Lemaître), we suspect that he thought of Friedmann's dynamic solutions as mathematical gimmicks, devoid of cosmological significance. This is borne out by the draft of his retraction, which is kept in the Einstein Archives. The published last sentence of the retraction reads: 'Es zeigt sich, daß die Feldgleichungen neben den statischen dynamische (d.h. mit der Zeitkoordinate veränderliche) zentrisch-symmetrische Lösungen für die Raumstruktur zulassen.' (It turns out that the field equations, besides the static solution, allow dynamic (that is, varying with the time coordinate) centrally symmetric solutions for the spatial structure.) In the draft, that sentence continues: 'denen eine physikalische Bedeutung kaum zuzuschreiben sein dürfte.' (to which a physical significance can hardly be ascribed.) Fortunately, Einstein deleted that verdict (Einstein 1923b).

8

Redshifts: How to reconcile Slipher and de Sitter?

When, in 1917, de Sitter published his cosmological model, he suspected Slipher's nebular redshifts to be of cosmological significance. In that year, Slipher published a note on 'Nebulae', where he cited their high redshifts in support of the 'island-universe' theory. When, in 1923, Eddington inserted Slipher's redshifts in tabular form into his influential textbook on General Relativity, he pointed to their potential as a highly significant observational clue to the structure of the Universe. His appeal was heeded. Whereas in 1922 Wirtz discussed nebular redshifts without referring to de Sitter, we see in 1924 and 1925, Wirtz, Silberstein, Lundmark and Strömberg trying to connect de Sitter's theoretical prediction with observations.

By that time, the theoretical debate on cosmology had become stuck in discussions on transformations of coordinate systems. Also, the observational side was struggling. Slipher's redshifts were of good quality, but when trying to find a redshift–distance relationship, they battled against the large uncertainties of nebular distances. Hubble twice delivered vital ammunition. The first time was in 1924/25 with the Cepheids in NGC 6822, M33 and M31. The second was contained in his 1926 publication of distances for over 40 nebulae. In 1927, Lemaître would be the first to reap the rewards gained from these improved distances.

Redshifts

The Lowell Observatory in Flagstaff, Arizona, was founded in 1894. The mysterious 'channels' on Mars, discovered by Schiaparelli in 1877, were the prime motive for constructing the site. In 1930, it was the place where Pluto was discovered. However, its most important discovery was the first message

from the expanding universe. In 1913, Vesto Slipher published in the Lowell
Observatory Bulletin 58 what he said were the first radial velocities (v_r) determined in spiral nebulae (Slipher 1913). Slipher had measured the wavelengths,
λ, in the spectra of nebulae. He saw that they were displaced by an amount, $\Delta\lambda$,
against the wavelength of the corresponding element when measured in the
laboratory. According to the Doppler shift formula, the displacement in wavelength could be attributed to a velocity, v_r, of the light source towards us (blue
shift) or away from us (red shift):

$$\frac{\Delta\lambda}{\lambda} = \frac{v_r}{c},$$

where c stands for the velocity of light. From September to December 1912, he
had taken spectra of the Andromeda nebula. He found $v_r = -300$ km/s. This was
the highest velocity for any astronomical object measured up to that time.
Slipher poses the question whether this *velocity-like displacement* might not be
due to some other cause, but then concludes that the Andromeda nebula is
indeed approaching the Solar System with a speed of 300 km/s. Slipher extended
his set. By 1915, he had observed line shifts of 15 nebulae: for the 11 he gave

Fig. 8.1 Slipher at the Lowell Observatory. Left: Vesto Slipher (1875–1969).
Right: The 24-inch Clark telescope with spectrograph. (Lowell Observatory
Archives.)

velocities, 2 of them were negative and 9 were positive (Slipher 1915). In 1917, he published radial velocities for 25 spiral nebulae: 4 of them show negative velocities, the other 21 move away from us (Slipher 1917).

Slipher was no great propagandist of his discoveries; it was Eddington who gave them wide publicity. In his book on General Relativity he inserted a list of 41 radial velocities of spiral nebulae taken by Slipher. The number of nebulae with negative velocities was still only 4, whereas the number of redshifted nebulae had increased to 37, their velocities ranging from +150 to +1800 km/s (Eddington 1924, p. 162). The growing number of strongly redshifted spectra strengthened suspicions about the cosmological connection to redshifts predicted in de Sitter's model.

Distances

Early interpretations of redshifts were hampered by totally inadequate distance determinations to extragalactic nebulae. In 1925, the debate about island universes had taken a decisive turn with Hubble's publication on the extragalactic nature of NGC 6822, M33 and M31. He followed it in 1926 with a statistical investigation of 400 extragalactic nebulae, where he correlated nebular diameters, d, with observed magnitudes, and with distances, D. He found: 'The coefficient of log (d) corresponds with the inverse-square law, which suggests that the nebulae are all of the same order of absolute luminosity and that apparent magnitudes are measures of distance.' (Hubble 1926, p. 321.) For the total magnitude, m_T, he used the formula $m_T = C - 5 \log (d)$. C is a parameter that depends on the type of galaxy. With the appropriate C, the nebulae can be reduced to a standard type. This leads to a single distance–magnitude formula. Hubble found the mean absolute visual magnitude, as derived from the nebulae whose distances were known, to be –15.2. His relation between magnitude and distance became $\log D = 4.04 + 0.2 m_T$, where m_T is the total apparent magnitude, and D is given in parsecs (Hubble 1926).

On the last three pages of his article, Hubble estimated the number of galaxies in a given volume. This gave him a mean density for the observable Universe of 1.5×10^{-31} g/cm^3. For the next few years this was the generally accepted number for the density of the Universe. To calculate the size and mass of the finite but boundless Universe, Hubble referred to the 1917 theory of de Sitter, but then employed a formula taken from Haas which refers to the Einstein universe (Haas 1925). As mentioned earlier, he found a radius of curvature of 2.7×10^{10} pc and a total mass of 9×10^{22} solar masses. According to Dr Sandage, the Mount Wilson library still has the Haas book with marginal notes in pencil by Hubble (Sandage, personal communication, 2007).

Early interpretation of redshifts

In 1922, Carl Wirtz published a short article entitled: 'Einiges zur Statistik der Radialbewegungen von Spiralnebeln und Kugelsternhaufen.' (Wirtz 1922.) (Remarks on the statistics of the radial motions of spiral nebulae and globular clusters.) It was a time when the distinction between galactic and extragalactic objects was not yet clearly established. Wirtz found that redshifts of spiral nebulae presented a totally different pattern from the redshifts of stars and globular clusters: the main motion of the nebulae could be described as a dispersal (Auseinandertreiben) relative to us, those nearer to us having a smaller recession velocity than those further away; this radial motion of the spiral nebulae had nothing to do with the motion of the Sun.

Two years later, Wirtz returned to the subject to discuss de Sitter's cosmological model in the light of Slipher's redshifts. He referred to Eddington's first edition of *The Mathematical Theory of Relativity* (Eddington 1923/1924) and recalled that in de Sitter's model, redshifts in remote objects are expected for two different reasons: (1) the flight of objects towards the particle horizon, and (2) a tendency for time to move slower in far away objects, even if they are stationary in relation to us. He then asked whether the slowing down process could be observed. For that purpose he collected apparent photographic diameters, Dm, of spiral nebulae from different sources and compared them with Slipher's velocities given in Eddington's book on relativity. From these data he derived the relation $v = 2200 - 1200 \log(Dm)$ (v in km/s, Dm in arcmins). This result, Wirtz says, is compatible with de Sitter's model (Wirtz 1924).

In 1924, Silberstein published a relation between Doppler shifts, $\Delta\lambda$, distances, r, and the curvature of the Universe, which he had derived from de Sitter's theory (Silberstein 1924, and several other publications in the same year). It has the form $\Delta\lambda/\lambda = \sin(r/R)$, where r is the distance to the object, and R the curvature of spacetime. For objects where r is small compared with R, this results in a linear relationship between the wavelength shift $\Delta\lambda$ and r:

$$\frac{\Delta\lambda}{\lambda} \approx \pm\frac{r}{R}.$$

Thus he could account for positive as well as negative line shifts. Lemaître (1925b) expressed doubts about the formula and, in a letter to *Nature* in May 1924, Eddington doubted the physical insight of Silberstein (Eddington 1924). Moreover, Silberstein had applied the formula to globular clusters in our Galaxy. This was, of course, doomed from the very outset. However, the attempt inspired extensive investigations by Lundmark (1925) and Strömberg (1925).

Nebular velocities were extracted from the equations determining the solar motion relative to nebulae:

$V = X \cos a \cos \delta + Y \sin a \cos \delta + Z \sin \delta + K$, with a and δ being right ascension and declination. In this expression, V is the observed radial velocity, X, Y, Z are rectangular equatorial components of the group motion: $-X = V_0 \cos A_0 \cos D_0$, $-Y = V_0 \sin A_0 \cos D_0$, $-Z = V_0 \sin D_0$. V_0, A_0, D_0 are the velocity and the coordinates of the solar apex relative to the group under consideration. The correction term K would take care of any other systematic motion. Thus, if the nebulae represent a motionless, fixed background, then X, Y, Z define the solar motion within that group, and the correction term K will vanish. However, if the nebulae have peculiar motions, then the correction factor will be $K \neq 0$.

'The determination of the curvature of space-time in de Sitter's world' was the title of Lundmark's 1924 publication, in which he enquired about the possibilities offered by open clusters, like the Pleiades, globular clusters, novae, various types of stars and spiral nebulae, to determine the curvature of spacetime, according to Silberstein's formulae given above (Lundmark 1924). We find in his paper the first 'Hubble diagram'. He plotted radial velocities against distances for globular clusters, Cepheids, O stars, eclipsing variables and also for spiral nebulae. For the spiral nebulae, he plotted velocities in km/s against distances in units of the distance to Andromeda, which he gave as 200 000 parsecs. The scatter was large, and he concluded on the negative note: 'we see that R cannot be determined at least from our present material with any accuracy.' In a follow-up publication, he stated that the best agreement was obtained with a correction term $K = k + lr + mr^2$, and: 'A rather definite correlation is shown between apparent dimensions and radial velocity, in the sense that the smaller and presumably more distant spirals have the higher space velocity.' (Lundmark 1925.) This was a transitional work. He estimated distances to several nebulae. Thus, from a study of novae, he found 1.4 Mly (million light years) for Andromeda – today's accepted value is 2 Mly. Whilst writing his paper, Lundmark learnt that Hubble had just found Cepheids in Andromeda, for which he gave a distance of 0.93 Mly (Hubble 1925b). Lundmark had no doubt about the extragalactic nature of some of the nebulae; his opinion was: 'Our stellar system and the system of spiral nebulae are constructed according to the conception expressed in the Lambert–Charlier cosmogony.'

The Charlier-Lambert cosmogony advocated an infinite, static, hierarchical and fractal universe. It distinguished galaxies of order 1, 2, 3, etc. With increasing order, galaxies are supposed to be larger, without containing more stars. Thus, each order diminishes the mean density of the Universe, which, in this way, can approach zero. This solves the Olbers paradox (Charlier 1922). However, this fractal universe proved to be incompatible with observations.

In an analysis of radial velocities of globular clusters and non-galactic nebulae in 1925, Strömberg had two goals in mind: (1) to check if the large solar velocity against these objects might provide a fundamental reference system, and (2) to ascertain whether the velocities gave any evidence of a curvature of spacetime according to the model of de Sitter. His velocity table contains mainly observations by Slipher. Strömberg was still hampered by the question of whether his nebulae and globular clusters might lie at the same distance. He found that observational uncertainties of nebular distances prevented any meaningful conclusions (Strömberg 1925).

That situation changed when Hubble published his long list of calibrated magnitudes and distances in 1926, and established the uniform distribution of nebulae in the observationally accessible part of the Universe (Hubble 1926). Finally, there existed a consistent set of distances to galaxies.

9

Lemaître discovers the expanding universe

'Un univers homogène de masse constante et de rayon croissant, rendant compte de la vitesse radiale des nébuleuses extra-galactiques.' (Lemaître 1927.) (A homogeneous universe of constant mass and increasing radius, accounting for the radial velocity of extragalactic nebulae.) This is the title by which, in June 1927, Lemaître announced his discovery of an expanding universe. Lemaître is unequivocal, for the first time someone suggests an expanding universe. This revolutionary statement was supported by a combination of theoretical and observational arguments.

Lemaître's education was very propitious for participating in the cosmological endeavour of the 1920s. He acquired a solid background in physics and mathematics early on, and was fortunate to have Eddington as a teacher, the foremost astrophysical theoretician of his time. Then he worked in Shapley's group, one of the most influential astronomers.

His independent research began in 1925 with an examination of de Sitter's model. Lemaître spotted its flaws without, however, offering a substantial alternative. That came in 1927. With a fresh look at Einstein's fundamental equations and incorporating the observational progress accomplished in those years, he found our Universe to be expanding. In 1927, this discovery was published in French in a journal that was not widely read. Only in 1930 did Eddington and de Sitter become aware that Lemaître had solved their most enigmatic cosmological problem.

Lemaître, a student of Eddington and research fellow in Shapley's group

Georges Lemaître was born 1894 in Charleroi (Belgium). Military service in World War I interrupted his university education. After the war he studied physics and mathematics. In 1920, he obtained a doctorate in mathematics (the Belgian doctorate was then of less academic value than after the university reforms of 1929). At the same time, he began a theological education and was ordained as a priest in September 1923. More details on his private life are given in the biography by Lambert (2000).

In a contest for a scholarship abroad, Lemaître had presented a thesis on Einstein's relativity. He succeeded and in 1923 became a graduate student in astronomy at the University of Cambridge. There he worked for one year with Eddington, who drew him further into modern cosmology. In Eddington's correspondence there is a letter dated 24 December 1924 to De Donder of the University of Brussels, where he wrote (Vibert Douglas 1956, p. 111): 'I found M. Lemaître a very brilliant student, wonderfully quick and clear sighted, and of great mathematical ability. He did some excellent work whilst here, which I hope he will publish soon. I hope he will do well with Shapley at Harvard. In case his name is considered for any post in Belgium I would be able to give him my strongest recommendation.'

Lemaître spent the academic year 1924/25 with Harlow Shapley at Harvard College Observatory in Cambridge, Massachusetts. Shapley was then still involved in nebular research. Lemaître arrived when the drawn-out debate about the existence of island universes was heading towards its close. At the same time, he registered at the Massachusetts Institute of Technology (MIT) for a doctorate in sciences. In 1924, the Harvard College Observatory (HCO) was not entitled to award Ph.D. degrees. According to Gingerich (personal communication 2007), the first official Ph.D. degree from HCO was awarded in 1929 to Frank Hogg. Cecilia Payne – later Payne-Gaposchkin – had earned her Ph.D. at HCO in 1925, but the title was awarded by the Radcliffe College, one of the Seven Sisters women's liberal-arts colleges. No such obstacles existed at MIT. Lemaître finished his thesis on 'The gravitational field in a fluid sphere of uniform invariant density according to the theory of relativity' in 1926, after returning to Belgium in 1925. There he taught at the University of Leuven (Louvain). MIT awarded him the Ph.D. in July 1927.

In the Harvard College Observatory Annual Report of the Director for 1 October 1924 – 30 September 1925 we find: 'Dr. Georges Lemaître, of the University of Louvain, has studied the problem of the pulsation theory of Cepheid variables while at the Harvard Observatory as a Fellow of the C.R.B. Educational Foundation. His work will be published in a forthcoming Circular.' Indeed, a 'Note on the theory of pulsating stars' by Lemaître was submitted on 10 May 1925 and appeared as Circular 282. In this work he looked for computational methods to simplify the comparison between different types of pulsating variables. There is a footnote, most probably added by

Shapley: 'Fr. G. Lemaître, Sc.D., of the University of Louvain, while a fellow of the C.R.B. Education Foundation, has spent the past year in the study of physics and astronomy at the Massachusetts Institute of Technology and the Harvard College Observatory. (HS.)'

Although Shapley was certainly interested in anything that might lead to a generalisation of the Cepheid period–luminosity relation, we find no indication of a close relationship between Shapley and Lemaître. In his autobiography, Shapley does not mention Lemaître (Shapley 1969).

In 1950, Lemaître reviewed a book, *L'expansion de l'univers*, by Paul Couderc. In this review he inserted the following biographical information about his status in 1927: 'I was … a member of the International Astronomical Union (Cambridge, 1925), and I had studied astronomy for two years, one year with Eddington, and another year in the American observatories. I had visited Slipher and Hubble and heard the latter in 1925 in Washington giving his memorable communication on the distance of the Andromeda nebula.' (Lemaître 1950, translated from French.) His memory must have played a trick on him because, as we have noted before, Hubble was not in Washington to present his 'memorable communication', it was read by Russell.

Doubts about de Sitter's choice of coordinates

With his education in theoretical cosmology and practical astronomy, Lemaître was well placed to embark on one of the central cosmological themes of those years: how could cosmological models account for the observed redshifts in spiral nebulae, and how is the de Sitter effect to be interpreted? The general cosmological debate revolved around two deficient models: that of Einstein, which was unable to explain redshifts, and the other of de Sitter, which contained no matter. Both were built on the assumption that our world was static. Lemaître would find the answers in 1927 by dropping the assumption of a static universe, working in an appropriate system of coordinates, and combining two sets of observations.

His first contribution appeared in 1925 (Lemaître 1925a,b). Its content was purely theoretical. He accepted the premises of General Relativity and a 4-dimensional universe of constant positive curvature. He opted for a division of spacetime into a 1-dimensional time and a 3-dimensional space. Then he spotted de Sitter's weak point. Although the model is homogeneous in spacetime, de Sitter's coordinate system offends the principle of homogeneity of space. His coordinates distinguish a centre, for which there is no counterpart in reality. Away from the central position of the observer, time is running slower. Thus, at a given time, clocks are running differently at different locations.

Spacewise, only the observer situated at the origin of the coordinate system moves on a geodesic. This grossly offends the principle of homogeneity.

Lemaître corrected de Sitter's inconsistency by choosing a different coordinate system, where the flow of time was the same at any location. In his representation, the parameter R, describing the curvature of space, was no longer a constant, as it had been in the solutions of Einstein and de Sitter, but was replaced by an expression that depended on time. His new choice of coordinates had two effects. The gravitational field is no longer static, and the 3-dimensional space is now Euclidean. Redshifts of distant objects, the great attraction of de Sitter's model, are also a natural feature of Lemaître's 1925 model.

Lemaître did not like the result. He comments: 'Our treatment evidences this non-statical character of de Sitter's world which gives a possible interpretation of the mean receding motion of spiral nebulae. The second point, on the contrary, seems completely inadmissible. We are led back to the Euclidean space and to the impossibility of filling up an infinite space with matter which cannot but be finite. De Sitter's solution has to be abandoned, not because it is non-static, but because it does not give a finite space without introducing an impossible boundary.' He obviously sensed that his approach might do away with de Sitter's paradoxes, yet retain the capacity to explain the observed nebular redshifts, but he was not prepared to accept the infinite Euclidean space.

Later it turned out that Lemaître's new solution was a special case among a wider choice of valid possibilities. Indeed, today's cosmology is based on curved spacetime, with a Euclidean (flat) spatial part, but this part is in accelerated or decelerated expansion, depending on the model. Thus, the 3-dimensional space is flat, but the 4-dimensional spacetime is curved. Examples for such configurations are shown in Chapter 6, 'The early cosmology of Einstein and de Sitter'.

Lemaître presented a paper to the April 1925 Washington meeting of the American Physical Society, the abstract of which is published in the *Physical Review*. His affiliation is given as Massachusetts Institute of Technology, where he was inscribed as Ph.D. student! The abstract (Lemaître 1925b) is really a summary of the above-mentioned publication; here is the full text:

> De Sitter's solution of the relativistic equations of the gravitational
> field in the absence of matter introduces a spurious inhomogeneity
> which is not simply the mathematical appearance of center of an origin
> of coordinates, but really attributes distinct absolute properties to a
> particular point. A new form of de Sitter's solution is given in order to
> remove this difficulty. New coordinates are introduced and, with the
> corresponding separation of space and time, the field is found
> homogeneous but non-statical. Furthermore the geometry is euclidean.

The singularity at de Sitter's horizon disappears. The Doppler effect has the numerical value given by Silberstein, but no way is found of introducing his double sign without spoiling the homogeneity of the field. The result that a homogeneous solution of de Sitter's world is non-statical and euclidean is rather against the physical significance of de Sitter's universe. However, the non-statical character of this universe has been advocated as an advantage by Eddington, but euclidean geometry is a very unsatisfactory feature of any conception of the whole universe.

The discovery of the expanding universe

His astronomical knowledge and mastery of General Relativity stood him in good stead again when he returned to the cosmological problem in 1927. In his publication, he first recapitulates the major characteristics of Einstein and de Sitter's models: the universe of de Sitter ignores the existence of matter, it is difficult to interpret, but it offers an explanation of the observed redshifts as consequences of the properties of the gravitational fields. Einstein's model takes the evident existence of matter into account and allows one to calculate the mass and extent of the Universe. It seems desirable to find an intermediate solution of the field equations combining the advantages of both. At first sight this seems impossible. But it can be done; one just has to realise that de Sitter's model does not meet all the requirements (Lemaître 1927, 1931a):

> Space is homogeneous with constant positive curvature; space-time is also homogeneous, for all events are perfectly equivalent. But the partition of space-time into space and time disturbs the homogeneity. The co-ordinates used introduce a centre. A particle at rest otherwise than at the centre does not describe a geodesic. The coordinates chosen destroy the homogeneity and produce the paradoxical results which appear at the so called "horizon" of the centre. When we use co-ordinates and a corresponding partition of space and time of such a kind as to preserve the homogeneity of the universe, the field is found to be no longer static; the universe becomes of the same form as that of Einstein, with a radius no longer constant but varying with the time according to a particular law.

In a footnote, Lemaître describes the difference between de Sitter's coordinate system and his own choice in the following way: 'In a representation restricted to two dimensions, one for time the other for space, the division of space and time used by de Sitter can be represented on a sphere: space lines are furnished

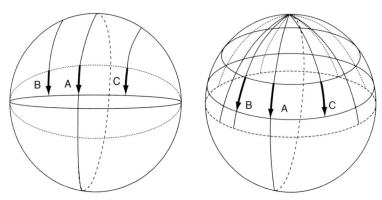

Fig. 9.1 Coordinate systems chosen by de Sitter and Lemaître. The 3-dimensional spatial part at a given moment in 'cosmic' time is projected into a 1-dimensional circle. Time runs from the north to the south pole. Left (de Sitter): In de Sitter's world, the spatial part is projected into a great circle. For different times they all cross at the same point, where 'time stands still'; two of them are shown. Point A moves on a geodesic (a great circle). Points B and C move on world-lines that run parallel to the great circle. However, they are not great circles and therefore not geodesics. This violates the principle of homogeneity. Time runs fastest for point A. It slows down with increasing distance from the geodesic line; time stands still at the two extremes. This causes the de Sitter redshift. Right (Lemaître): The spatial line is a circle of latitude; the temporal line is a meridian. The points A, B and C move on geodesics. Their separation increases as time progresses. Time runs in the same way for all points.

by a system of great circles, crossing on the same diameter, and the temporal lines are parallels crossing at right angles the space lines. One of these parallels is a great circle and therefore a geodesic, it corresponds to the centre of space, the pole of this grand circle is a singular point, corresponding to the horizon of the centre. Naturally, this representation has to be extended to four dimensions, and the time coordinate has to be taken as imaginary, but the lack of homogeneity resulting from the particular choice of coordinates remains. The coordinates respecting homogeneity correspond to taking the temporal lines as the meridians, and as spatial lines the latitude parallels, in that case the radius of space varies with time.' (Lemaître 1927, translated from French.) This footnote was not given in Lemaître (1931a).

Figure 9.1 is a simplified drawing that clearly shows the difference between de Sitter and Lemaître's choice of coordinates. Lemaître's world models will be further discussed below and in Chapter 17, 'The seed for the Big Bang'.

Lemaître set out to find a solution of Einstein's fundamental equations, where the radius of the Universe is allowed to vary in an arbitrary way. In addition, the coordinates have to respect the homogeneity of space, and at any given moment, time has to run in the same way for any point in space. He treats the Universe as a rarefied, homogeneous gas; molecules represent the galaxies. Pressure, being two thirds of the kinetic energy, is negligible with respect to the energy associated with matter. However, he keeps the pressure term in the fundamental equations; after all, radiation pressure might have to be taken into account. Indeed, radiation was eventually going to play a crucial role in Lemaître's early universe.

Lemaître's solutions of the fundamental equations of relativity correspond to Einstein's closed, curved universe. The decisive difference was that the radius of curvature R – the radius of the Universe – was allowed to change in time. The constant R becomes a function that varies with time: $R = R(t)$. Lemaître's

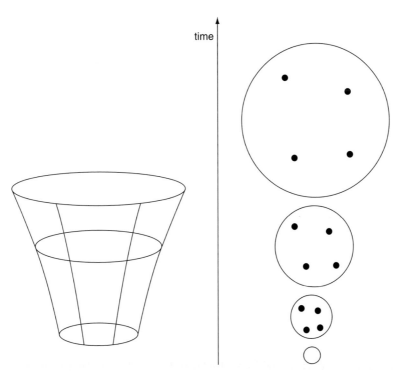

Fig. 9.2 Lemaître's expanding universe. We show two ways of representing the expanding universe. Left: The 3-dimensional curved spatial universe is projected into one dimension on a circle (that universe is spatially closed). Time evolves upwards along the vertical lines; a galaxy at rest moves along such a line. The radius of curvature of the 3-dimensional space increases with time. This is mirrored in the increasing radius of the circle. Its evolution in time forms a hyperboloid. Right: The 3-dimensional curved spatial part of the universe is projected onto a 2-dimensional sphere. Due to the expansion, the radius of the sphere increases.

universe shrinks or grows. From observations, he convinced himself that our Universe is expanding: $R(t)$ is growing with time. Lemaître's universe can be represented as a hyperboloid. The horizontally placed circles represent the 3-dimensional spatial world projected into a 1-dimensional circle. The circle shows that the 3-dimensional space is closed. The motion of a point in spacetime can be represented on a hyperboloid. A point moving along a vertical line is spatially at rest. This universe has a curved spacetime: Lemaître's world is dynamical.

Lemaître kept the cosmological term, Λ, in his model. Einstein had introduced it in order to obtain a static universe. But, in 1931, when Einstein finally accepted the expanding universe, he evicted Λ from further deliberations. Lemaître differed fundamentally. He saw in it much more than a stopgap. Lemaître and Eddington considered Λ played an essential part in the history of the Universe.

The derivation of the linear velocity–distance relationship

Redshifts in the spectra of distant spiral nebulae were a natural consequence of Lemaître's model. The expanding universe not only lengthens distances between galaxies, it also changes the metric of the 4-dimensional spacetime. Before discussing this result, we recall the various interpretations of the redshift of a spectral feature with an original wavelength λ by an amount $\Delta\lambda$.

In earlier chapters we mentioned that the classical Doppler effect explains a redshift by a fraction $\Delta\lambda / \lambda$ as the result of a movement of the emitting object away from the observer with velocity $v = (\Delta\lambda / \lambda) \times c$, where c stands for the velocity of light. Thus, a high redshift indicates a high velocity. With that interpretation, most of Slipher's spiral nebulae move away from us with average velocities of several hundred kilometres per second. From de Sitter's theory, two kinds of redshifts were expected. A test particle placed close to the observer is accelerated away from the observer. The resulting velocity produces a classical Doppler shift. In addition, a test particle at some distance from the observer emits a redshifted spectrum, even if the particle remains at rest because, in de Sitter's static universe, time is slowing down with increasing distance from the observer.

Lemaître's hope of explaining the nebular redshifts became true for a different reason. During expansion, the radius of curvature increases, which results in a change of the metric. This produces a redshift in the light that crosses the expanding universe. If an atom emits a photon with wavelength λ_e at the time when the radius of curvature is R_e, and an observer measures that same photon at a later time when the radius of curvature has grown to R_a, then the wavelength of that photon has lengthened to λ_a. The wavelength grows in the same sense as the Universe, such that the ratio of the absorbed wavelength to the emitted wavelength is the same as the ratio of the radii of curvature (the sizes of the Universe), thus: $\lambda_a/\lambda_e = R_a/R_e$. This insight led Lemaître directly to the famous velocity–distance relationship, which is now called the 'Hubble relationship'. For Lemaître's derivation of this and the following relations, see the Mathematical Appendix.

When representing the closed universe as a sphere, we see immediately that the distance, r, between two galaxies is growing as the size of the sphere increases; it will grow in the same sense as the radius of the Universe. The light from a distant galaxy will take longer to reach us than the light from a close galaxy. If we live in a steadily growing universe, the ratio R_a/R_e will be larger for a photon that has travelled a long way. We therefore expect the redshift to increase proportionally to the distance, r, between emitter and receiver: $\Delta\lambda/\lambda = k \times r$. The exact amount of the shift per unit distance, k, has to be determined by theory or observations.

Lemaître followed the habit of expressing redshifts in velocities, according to the Doppler formula $\Delta\lambda/\lambda = v/c$. If we equate this Doppler formula with Lemaître's redshift–distance relationship – $\Delta\lambda/\lambda = k \times r$ – we find $v/c = k \times r$, or $v = c \times k \times r$, which we abbreviate to the famous formula, $\boldsymbol{v = H \times r}$.

Lemaître derived this linear velocity–distance relationship theoretically in 1927, and it is now called the 'Hubble relationship'. In the same publication, Lemaître also calculated H, the constant of proportionality, now known as the 'Hubble constant'.

The velocity–distance relationship presents the observed redshifts as if they are caused by the velocity of the observed galaxy. The 'Hubble constant', H, tells us how much that velocity increases, measured in km/s, when the distance increases by 1 Mpc. But Lemaître cautions against that interpretation. The observed galaxy may have a peculiar motion independent of the cosmological expansion, which will result in an ordinary Doppler shift, and that can be a blue shift or a redshift. However, the peculiar motion will diminish in importance relative to the speed of separation due to expansion. The cause of the cosmological redshift is not the relative velocity, but the change of the metric between the emission and absorption of the photon.

Indeed, in today's cosmological discussions, the cosmological redshift is mostly given as z, where z is defined as $z = \Delta\lambda/\lambda$. It avoids the fallacious association of the cosmological redshift with a Doppler shift. A simple mathematical reformulation, with the help of $\lambda_a/\lambda_e = R_a/R_e$, leads to $z = R_a/R_e - 1$. Thus, z connects the redshift with the change in the size of the Universe between the emission and absorption of the photon.

Today, we see $R(t)$ as the scale factor indicating the temporal variation of the expansion, rather than the parameter of curvature. We are more interested in the history of $R(t)$ than its absolute value, and in related questions, like which forces were and are active in accelerating and braking the expansion of the Universe.

Lemaître determines the 'Hubble constant' from observations

Lemaître then combined theory and observation. His stay at Harvard and his visits to observatories and scientific meetings had familiarised him with the observational side of cosmology. Thus, he knew where to find the observational corroboration or falsification for his theoretical model. He had a set of 42 extragalactic nebulae, for which he essentially matched Slipher's red-shifts against Hubble's distances. He referred to the radial velocities tabulated in Strömberg (1925), and the magnitudes in Hubble (1926). He converted Hubble's apparent magnitudes m into distances, employing Hubble's formula for the relation between m and distance r: $\log r = 0.2m + 4.04$ (Hubble 1926). The uncertainties in the individual observations, particularly in the distances, were very high. However, the positive correlation between redshifts and distances was obvious, and Lemaître was satisfied that at some later date more accurate observations would confirm beyond doubt the theoretically expected linear velocity–distance relationship.

Let us now look at Lemaître's calculation of the Hubble constant. Because of the large uncertainties in the individual observations, Lemaître decided to compare the mean distance of his 42 galaxies with their mean velocity. Assuming a solar motion of 300 km/s in the direction of $\alpha = 315°$, $\delta = 62°$, he found for the chosen sample a mean distance of $r = 0.95$ Mpc and a mean velocity of $v = 600$ km/s. To take account of the greater uncertainties of the more distant objects, he at first weighted the observations according to $1/\sqrt{1 + r^2}$ (r measured in Mpc). Lemaître then derived the Hubble constant, $H = v/r$, which he calculates to be 625 (km/s)/Mpc. There is a slight inconsistency in that number. With his values for v and r, we obtain $H = 632$ (km/s)/Mpc, so he must have done some rounding somewhere. Anyway, he repeats the calculation by giving the same weight to all galaxies, and now obtains $H = 575$ (km/s)/Mpc. We have here the first determination of the Hubble constant. This section, where Lemaître calculated distances and extracted the Hubble constant, was for unknown reasons omitted in the translation of Lemaître's 1927 publication, which appeared in the *Monthly Notices of the Royal Astronomical Society* of 1931 (Lemaître 1931a). We print the omitted section in Chapter 11, 'The breakthrough for the expanding universe'.

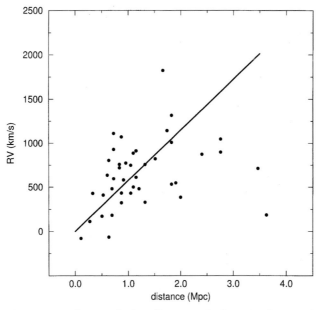

Fig. 9.3 Lemaître's velocity–distance relation. The diagram shows the observations that helped Lemaître justify his theoretically found velocity–distance relationship and calculate the 'Hubble constant': radial velocities (RV) against distance. Lemaître did not publish this diagram, but he gave the sources of his data. Duerbeck and Seitter (2000) have reconstructed the points. From these observations Lemaître derived $H = 575$ (km/s)/Mpc; the corresponding line has been inserted by H. Duerbeck (Duerbeck 2008, personal communication). Individual error bars would be small for velocities but large for distances. Records for drawing error bars are lacking. (Duerbeck and Seitter 2000, and Duerbeck 2008.)

In a footnote, Lemaître comments that certain authors have tried to find a relationship between v and r, but detected only a feeble correlation between them. Considering the badly determined distances, this does not astonish him. He concludes that the available observations made it possible only to assume v is proportional to r, and to avoid a systematic error in the determination of the ratio v/r. A first step in fulfilling Lemaître's hope for an observationally irreproachable confirmation was realised two years later by Hubble (Hubble 1929a).

Lemaître's interpretation of theory and observations

Lemaître understood the expanding universe as we understand it today. On 31 January 1929, more than one year before the international scientific community took note of the expanding universe, Lemaître explained how the

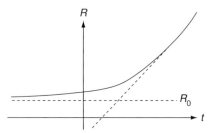

Fig. 9.4 The early Universe: Lemaître's version of 1927. The expanding universe
emerges out of a quasi-static Einstein universe with radius R_0. This was also
Eddington's favourite version. In 1931, Lemaître will propose a universe being born
or re-born out of a primeval atom: this became our Big Bang scenario, which is
discussed in Chapter 17.

nebular redshifts have to be understood at a conference to the Société scientifique de Bruxelles (Lemaître 1929). He employed the same pictures we use today:
the nebulae remain in the same configuration in space, but space itself increases
its size in the course of time. Thus, the distance between two nebulae covers the
same fraction of space, but grows together with space. Therefore any two
nebulae draw away from each other. They are like microbes on a bubble.
When the bubble increases, each microbe realises that the others withdraw,
and it has the impression – but only the impression – of being at the centre.

What about the beginning of the Universe? We shall dedicate a special
chapter to that question, but for now we restrict ourselves to presenting the
view Lemaître expressed in his 1927 publication. From a mathematical point
of view there are several possibilities. Lemaître opted for a solution where, in
the course of time, the radius of the Universe steadily increases from an initial
size, which corresponds to some equilibrium value. Or, to use Lemaître's formulation: 'The radius of the universe increases without limit from an asymptotic value R_0 for $t=-\infty$.' From his theoretical solution of Einstein's equations
and expansion rate of 625 (km/s)/Mpc, he calculated this asymptotic size to be
$R_0 \approx R_\varepsilon / 100$. R_ε represents the 'radius' of Einstein's static universe, extracted
from the observationally derived density, from which Hubble had in 1926
obtained $R_\varepsilon = 2.7 \times 10^{10}$ parsec. Thus, Lemaître's universe of 1927 does not originate out of a singularity and has no abrupt beginning: it expands towards de
Sitter's universe of vanishing density.

Lemaître's debt to Friedmann

When Lemaître's article was translated and published in the *Monthly
Notices* in 1931, he added several references that were not given in the original,

and one of them was to Friedmann (Lemaître 1931a). This might have conveyed the impression that he knew about Friedmann when writing his original work. However, when submitting his article in 1927, Lemaître was not aware of Friedmann's publications. He learnt about that pioneering investigation from Einstein on the occasion of the Solvay Congress in October 1927. In a footnote to his 1929 public lecture, Lemaître thanks Einstein for having directed his attention to the work of Friedmann, which, he says, already contains several notions and results later rediscovered by himself (Lemaître 1929). Thus, Lemaître owes nothing to Friedmann.

The two pioneers did not have the same priorities. Whereas Friedmann examined the different solutions of Einstein's field equations due to variations in the parameters ρ and Λ, Lemaître emphasised the history of the dynamical spherical universe. This culminated in his hypothesis of the primeval atom: the Big Bang – more about that later. And, whereas Friedmann was unaware of Slipher's redshifts, and therefore remained within the theoretical realm, Lemaître was very much aware of the newest astronomical observations and made full use of them.

Although Lemaître rediscovered the dynamical universe completely independently of Friedmann, he always acknowledged that Friedmann had been the first to find dynamical solutions.

Notwithstanding his ignorance of Friedmann, Lemaître was aware of some earlier critiques or comments on de Sitter's model. Thus, in 1927 he cites Lanczos (1922) as well as Weyl (1923c), and Du Val (1924) who had published a geometrical description of de Sitter's model. Actually, in his three papers of 1925 and 1927, he cites publications of Lanczos (1922), Eddington (1923), Weyl (1923c), Du Val (1924), Lundmark (1924), Strömberg (1925) and Hubble (1926). Without giving particular references, he also cites Silberstein's work of 1924, and, of course, the classical 1917 publications of Einstein and de Sitter.

Einstein judges Lemaître's interpretation as 'abominable'

In 1957, on the occasion of the second anniversary of Einstein's death, Lemaître talked on the Belgian radio about his meetings with Einstein (Lemaître 1958). Their first encounter occurred when Einstein participated in the Solvay Congress in the autumn of 1927. The already legendary Einstein and the young scientist discussed Lemaître's article. 'Après quelques remarques techniques favorables, il conclut en disant que du point de vue physique cela lui paraissait tout à fait abominable.' (After a few favourable technical remarks he concluded by saying that from a physical point of view this looked to him abominable.) During the taxi ride to Auguste Piccard's laboratory, Lemaître told Einstein

Fig. 9.5 Lemaître and Einstein. Georges Lemaître (1894–1966) and Albert Einstein (1879–1955), photographed around 1933. (Archives Lemaître, Université Catholique, Louvain.)

about the nebular velocities. He came away with the impression that Einstein was not at all well informed about astronomical facts. It is very likely that it was on this occasion that Lemaître learnt from Einstein about Friedmann's earlier venture into cosmology.

Einstein's conversation with Lemaître reminds us of his reaction to Friedmann. He first refused the dynamical universe for mathematical reasons. When he recognised that this reproach was untenable, Einstein retracted. However, he did not add a single word about the possible physical significance of Friedmann's work. Now, in 1927, Lemaître not only showed him a mathematically correct solution but, in addition, told him about the astronomical evidence. Yet, to Einstein an expanding universe, even when presented as a possibility emerging from his own theory, was simply 'abominable'. To convert him, it needed the heavyweights Eddington and de Sitter, and perhaps a further push by Tolman in January 1931; more about that later on.

In 1927, Lemaître had published his discovery in French in *Annales de la Société scientifique de Bruxelles*. None of the important figures in cosmology took any notice and the principal players, such as Eddington and de Sitter, continued racking their brains about the de Sitter effect.

Let us summarise Lemaître's achievements of 1927: he found the snag in de Sitter's model that had fooled cosmologists for ten years. He then went back to Einstein's fundamental equations of General Relativity and extracted a dynamical solution for the Universe. From this solution he derived a law for wavelength shifts in the spectra of remote light sources. Expressed in the terminology of Doppler shifts it is the famous linear velocity–distance relationship $v = H \times r$. He examined the relevant observations. They convinced him that we live in an expanding universe. From the observations, he also extracted the Hubble constant, H. Combining theory and observations he realised that the apparent 'moving apart' of the extragalactic nebulae is a cosmological effect caused by the expansion of the Universe.

10

Hubble's contribution of 1929

'About a year ago Mr. Hubble suggested that a selected list of fainter and more distant extragalactic nebulae, especially those occurring in groups, be observed to determine, if possible, whether the absorption lines in these objects show large displacements toward longer wave-lengths, as might be expected on de Sitter's theory of curved space-time.' Humason wrote this in 1929 as an introduction to a short note of one and a half pages, where he announced that spectrograms of NGC 7619, obtained with the 100-inch telescope, gave a radial velocity of 3779 km/s (Humason 1929). This was twice as large as any speed hitherto measured by Slipher. After having given further observational details, he points to a paper where Hubble 'gives approximate distances for 24 extra-galactic nebulae, and finds a marked increase in radial velocity with distance. The high velocity for NGC 7619 derived from these plates falls on the extra-polated line which expresses the relationship between line displacement and distance.' This refers to Hubble's famous announcement of a roughly linear relation between velocities and distances among nebulae. Hubble's paper appeared adjacent to Humason's. Humason continues: 'These results suggest an influence of distance upon the observed line shift – such as would be produced, for example, on de Sitter's theory, both by the apparent slowing-down of light vibrations with distance and by a real tendency of material bodies to scatter in space.'

Humason was a highly gifted and highly esteemed observer. Although he remained in Hubble's shadow, he played a crucial part in confirming the linear velocity–distance relation. He provided Hubble with observed redshifts from 1924 to 1953. Many of them came from very difficult spectroscopic observations because of the very low surface brightness of the galaxies. Humason's qualities must have been remarkable in many ways. According to Sandage (2004, p. 496),

'Humason was a superlative mule driver, fisherman, imprecationist, drinker, poker player, raconteur, rake and rogue, gentleman and friend.'

Hubble finds the linear velocity-distance relationship from observations

'Determinations óf the motion of the sun with respect to the extra-galactic nebulae have involved a term of several hundred kilometers which appears to be variable. Explanations of this paradox have been sought in a correlation between apparent radial velocities and distances, but so far the results have not been convincing. The present paper is a re-examination of the question, based on only those nebular distances which are believed to be fairly reliable.' (Hubble 1929a.) This is the introduction to Hubble's famous paper where he extracted the linear velocity–distance relationship from obser-vations. From the introductionary remarks we gather that the motion of the Sun in respect to extragalactic nebulae was the initial target of the paper. We shall see that when he arrived at his conclusions, the emphasis had changed.

Hubble had velocities (redshifts) to 46 nebulae, but only had distances for 24 of them. He obtained the seven most reliable distances from investigations on many individual stars. He derived another thirteen distances by assuming a uniform upper limit of stellar luminosities. The last four belong to the Virgo cluster, the distance to which he obtained from the distribution of nebular luminosities, together with luminosities of some of the stars. He also calculated the mean total visual magnitude for these 24 nebulae. He then compared the redshifts – most of them from Slipher – with the distances. The data indicated a linear correlation, irrespective of whether or not solar motion was considered. He then introduced the distances as coefficients of the K term. He saw no advantage in retaining the constant k in Lundmark's K term $(k+lr+mr^2)$, but he did not explicitly mention the mr^2 term (see the 'Early interpretation of redshifts' in Chapter 8). He chose a modified K term because he felt, at best, the data allowed a linear interpolation. He retained the lr term, which he calls rK, and thus his equation for solar motion took the form $v = rK + X \cos \alpha \cos \delta + Y \sin \alpha \cos \delta + Z \sin \delta$.

Depending on whether he gave equal weight to all the nebulae, or combined them into groups, Hubble obtained $K = 465$, or $K = 513$ (km/s)/Mpc. He concluded: 'The data in the table indicate a linear correlation between distances and velocities, whether the latter are used directly or corrected for Solar motion, according to the older solutions.'

For the group of 22 nebulae for which he could not determine individual distances, Hubble estimated a mean distance from the mean apparent magnitudes.

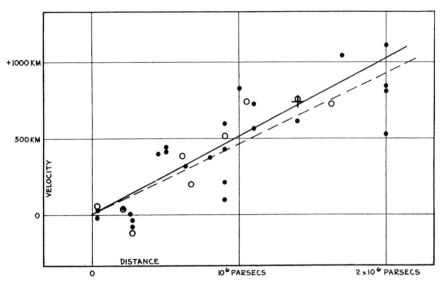

Fig. 10.1 Hubble's diagram of 1929. The velocity–distance relation, as published by Hubble in 1929. The full line corresponds to $H = 530 \,(\text{km/s})/\text{Mpc}$. The broken line results from a different weighting of the observations. (Figure 1 from Hubble 1929a.)

Comparing this mean distance with the mean of the velocities gave him $K = 530 \,(\text{km/s})/\text{Mpc}$. By the same method, Lemaître had arrived at $575 \,(\text{km/s})/\text{Mpc}$ in 1927. Since the underlying observational data show a large overlap, the close agreement should be expected.

Hubble felt certain about the nebular distances that covered the velocity range up to 1000 km/s. After discussing the quality of his data, he wrote: 'The results establish a roughly linear relation between velocities and distances among nebulae for which velocities have been previously published, and the relation appears to dominate the distribution of velocities. In order to investigate the matter on a much larger scale, Mr. Humason at Mount Wilson has initiated a program of determining velocities of the most distant nebulae that can be observed with confidence.' Hubble then referred to what for him must have been important additional evidence. Humason had found the largest redshift measured up to that time. We have mentioned this observation and Humason's comments in the introduction to this chapter. Humason found a radial velocity for NGC 7619 of $v = +3779$ km/s, or $+3910$ km/s when corrected for solar motion. This velocity nearly quadrupled the scale of Hubble's diagram, which stopped at 1090 km/s. From the cluster of which the nebula appeared to be a member, a distance of the order of 7 Mpc was estimated. That new observation resulted in $K = 500$. This obviously greatly strengthened Hubble's conviction that there was a direct link between distance and redshift, and that a linear relation between velocity and distance was a very good first approximation; it might even be a definite result.

Hubble's interpretation of the velocity–distance relationship

In 1929 Hubble did not spend much time on interpretation. He thought that new data expected in the near future might modify the results, and it would be premature to try to interpret them at any length. He presumed that it would become possible to find the motion of the galactic system with respect to the extragalactic nebulae. But then he added: 'The outstanding feature, however, is the possibility that the velocity–distance relation may represent the de Sitter effect, and hence that numerical data may be introduced into discussions of the general curvature of space. In the de Sitter cosmology, displacements of the spectra arise from two sources, an apparent slowing down of atomic vibrations and a general tendency of material particles to scatter. The latter introduces an element of time. The relative importance of these two effects should determine the form of the relation between distances and observed velocities; and in this connection it may be emphasized that the linear relation found in the present discussion is a first approximation representing a restricted range in distance.' These points had already been discussed in Eddington's well-known textbook (Eddington 1924, p. 161), and some of them in very similar words.

In 1929, Hubble did not consider the linear velocity–distance relationship as being definitely established. He knew about de Sitter's prediction of wavelength shifts: the slowing down of time resulted in a quadratic dependence on distance, whereas the scattering tendency of material bodies gave a linear dependence. Considering the scatter in the data, Hubble probably did not want to commit himself until more observations became available.

What was Hubble's motive for his 1929 investigation?

Hubble left no answer to that question, so we shall never definitely know. However, there are a few facts that lend themselves to conjecture. From the paper's introduction, which we cited in full at the beginning of the first section on p. 115, we learn that Hubble's 1929 investigation was originally meant to study solar motion relative to the nebulae. Hubble found that the velocity residuals correlated well with the nebular distances; he showed this in his famous diagram. Hubble was no theoretician, but from Humason's paper we know that by 1928 he was aware of de Sitter's prediction of large displacements towards longer wavelengths for distant objects. Hubble himself mentioned that de Sitter predicted two kinds of redshift, and said that their relative importance should determine the form of the relationship between distances and observed velocities; the observationally determined linear relationship had to be seen as a first approximation.

In 1928, Robertson had published a linear velocity–distance relationship (see Chapter 13, 'Robertson and Tolman join the game'). When the article appeared, Robertson was at Caltech. Sandage comments on the obvious question, did Hubble and Robertson talk about the velocity–distance relationship? In 1961, Robertson remembered having discussed his prediction with Hubble in 1928. However, during the four years that Sandage spent as Hubble's assistant (1949–53), the latter never mentioned that he had any knowledge of Robertson's work (Sandage 2004, p. 501).

It might well be that Robertson's explanations were too obscure for Hubble, and that he forgot about them. From all the information available we can certainly agree with Sandage that Hubble did not set out to verify any theory, nor to find some velocity–redshift relationship. He was after solar motion, but when he was blessed with Humason's large redshift for NGC 7619, he realised that he was on to something fundamentally new (Sandage 2004, p. 502).

The reception of Hubble's discovery

The first publications to cite Hubble's discovery confirmed that his results were regarded as highly significant. However, contrary to the impression conveyed by many writers of modern texts, his article was not seen as a stunning revelation. A relationship of that kind had been expected from the usual interpretation of de Sitter's theory, and some observers had been unsuccessfully looking for it. But with Hubble's results there could hardly be any doubt that such a relationship did exist, and it was very likely of a linear nature.

We find the first reference to Hubble's publication in Tolman's long treatise on the de Sitter line element (Tolman 1929b). Hubble and Humason had communicated their papers for publication on 17 January 1929, and Tolman submitted his paper on 25 February 1929. In the introduction, he said: 'The correlation between distance and apparent radial velocity for the extragalactic nebulae obtained by Hubble, and the recent measurement of the Doppler effect for a very distant nebula made by Humason at the Mount Wilson Observatory, make it desirable to consider once more the theoretical relations between distance and Doppler effect which could be expected from the form of line element for the universe proposed by De Sitter.' (Tolman 1929b, p. 246.)

The second reference comes from Zwicky, who submitted an article on spectral redshifts on 26 August 1929, from the same Pasadena address as Tolman. Zwicky writes: 'E. Hubble has shown recently that the correlation between the apparent velocity of recession and the distance is roughly linear ... Large deviations occur for the nearest nebulae, which may be attributed to their peculiar motions.' (Zwicky 1929, p. 773.) He then refers to more recent observations by Humason at very large distances, with very large deviations that could hardly be due to peculiar motions. Zwicky acknowledges that he obtained this information from personal discussions with Humason and Hubble.

Hubble's publication must have had quite an impact on de Sitter. We shall come back to this point in Chapter 12, 'Hubble's anger about de Sitter'. Our preliminary impression is that de Sitter saw the article, immediately realised its importance, and wanted to verify this theoretically significant observational relationship. He did, and found the same result (de Sitter 1930b).

Hubble was the first person to have collected sufficient observational evidence to propose a linear velocity–distance relationship based purely on observations. Again, based purely on observations, he calculated K, the proportionality factor between distance and velocity. Because these results were based uniquely on observations, they set landmarks that could not be disregarded, and served as guidelines in subsequent theoretical cosmological discussions.

Hubble and the expansion of the Universe

When the discovery of the expanding universe is attributed to Hubble, his 1929 publication is often quoted (if the statement is not simply copied from another secondary source). However, nowhere in those six pages does Hubble propose, nor even mention an expanding universe. He was aware that de Sitter had predicted a distance–velocity relationship that could be linear or quadratic. But de Sitter's prediction had nothing to do with an expanding universe. On the contrary, de Sitter's universe was meant to be static. In addition, Hubble never claimed to have discovered the expanding universe; he probably never believed in such a scenario.

We saw that Zwicky, as well as Tolman – the first to refer in print to Hubble's discovery – were intrigued by the observational results, but neither of them relates the new observational facts to an expanding universe. It should be remembered that at the time of Hubble's publication, no one in the astro-cosmological community talked about or believed in such a scenario - except Lemaître. That situation only changed in the middle of 1930, approximately one and a half years after Hubble discovered the velocity–distance relationship. More about this in the next chapter.

Leaflet 23 of the Astronomical Society of the Pacific, which appeared in July 1929, contains an excerpt from an article by Hubble with the title 'A clue to the structure of the universe'. At the end of the article he says: 'It is difficult to believe that the velocities are real; that all matter is actually scattering away from our region of space. It is easier to suppose that the light-waves are lengthened and the lines of the spectra are shifted to the red, as though the objects were receding, by some property of space or by forces acting on the light during its long journey to the Earth.' (Hubble 1929b.) Hubble was most probably thinking of Zwicky, whom he often met, and who at that time was speculating without success about an alternative to the de Sitter universe (Zwicky 1929).

The closest Hubble may have come to admitting that the velocity–distance relationship had something to do with an expanding universe was in his George Darwin Lecture on 8 May 1953, a few months before he died. There he said: 'Thus, if red-shifts do measure the expansion of the universe, we may be able to gather reliable information over a quarter of its history since expansion began, and some information over nearly half of the history.' (Hubble 1953.) However, he had made it clear at the beginning of his lecture that there were competing interpretations, and the right one was not yet known. That was in 1953, twenty-four years after his publication of the linear velocity–distance relationship.

The question as to why Hubble was elevated to the status of discoverer of the expanding universe belongs to sociology – and to the effect of public relations, through the media, on rewriting history. Kragh and Smith have devoted a very pertinent study to this phenomenon (Kragh and Smith 2003).

11

The breakthrough for the expanding universe

The Friday, 10 January 1930 meeting of the Royal Astronomical Society

On Friday, 10 January 1930, de Sitter gave a talk about the relationship between the velocities of spirals and their distances at the London meeting of the Royal Astronomical Society. It seems to have been a kind of preview of a paper he was going to publish in May of the same year (de Sitter 1930b). Commenting on a slide illustrating that relationship, he said: 'You see that the correlation is represented by a straight line, but you will naturally enquire how the distances are obtained. The distances of spirals can be estimated either from angular diameters, assuming constancy of linear diameters, or from their apparent integrated magnitudes, assuming constancy of the absolute integrated magnitudes. The two sets of data do not fit very well, but I have obtained formulae giving the distance from either the diameters or magnitudes and adjusted them so as to get reasonable estimates. In getting the constants of the formulae I made use of the distances obtained by Hubble, Lundmark, and Shapley.' The outcome was the same linear relation that Hubble had found. He admits that these observations are difficult to fit into his own model and into that of Einstein. In the ensuing discussion, Eddington refers to the solutions A and B (Einstein's and de Sitter's models) and says: 'One puzzling question is why there should be only two solutions. I suppose the trouble is that people look for static solutions. Solution A is such a static solution. Solution B is, on the contrary, non-static and expanding, but as there isn't any matter in it that does not matter.' Upon which de Sitter replies: 'It would be desirable to know what happens when we insert matter into the empty world represented by solution B. The difficulty in the investigation of this problem lies in the fact that it is not static.'

The proceedings of the meeting were published in *The Observatory* (de Sitter 1930a). Lemaître read the report. The Archives Lemaître at the Université Catholique de Louvain preserve a draft of a letter to Eddington, which is not always easy to decipher accurately: 'Dear Professor Eddington, I just read the February N° of the Observatory and your suggestion of investigating non statical intermediary solution between those of Einstein and de Sitter. I made these investigations two years ago. I consider an universe of curvature constant in space but increasing with time. And I emphasize the existence of a solution in which the motion of the nebulae is always a receding one from time minus infinity to plus infinity. This solved the question put forwards by de Sitter why the nebulae are on the receding branch of the hyperbola.' Lemaître then goes into technical details. Among other inferences, he points out that due to the redshift of the light, ghost nebulae cannot be observed. At the end of the letter he says: 'I send you a few copies of the paper. Perhaps you may find occasion to give it to de Sitter. I sent him also at the time but probably he didn't read it.' Lemaître then tells the story about Einstein's reaction. We heard it before when Lemaître reminisced on the occasion of the second anniversary of Einstein's death: 'I had occasion to speak in the matter wit Einstein two years ago. He told me that the theory was right [...?], that it was not new but had be considered by Friedmann, he made critic against which he was obliged to withdraw but that from the physical point vu it was 'tout a fait abominable'. Il hope you will not think so and I would be very pleased to year of your view. With kind regard I am yours sincerely' (Lemaître 1930a; we left grammar and spelling as we deciphered it.)

Thus, when he received news of Lemaître's discovery in 1927, Eddington had failed to realise its far-reaching consequences. McVittie, who in 1930 was a research student of Eddington's, relates in his obituary for Lemaître: 'I well remember the day when Eddington, rather shamefacedly, showed me a letter from Lemaître which reminded Eddington of the solution to the problem which Lemaître had already given. Eddington confessed that, though he had seen Lemaître's paper in 1927, he had completely forgotten about it until that moment.' (McVittie 1967.)

Lemaître's draft is not dated, but was obviously written some time in February or March 1930. Eddington's public reaction to Lemaître's message appeared in the May 1930 *Monthly Notices*, and that of de Sitter in the May 1930 *Bulletin of the Astronomical Institutes of the Netherlands*. Both Eddington and de Sitter acknowledged that Lemaître's paper was a revelation that made the scales fall from their eyes. They enthusiastically spread the gospel, and the concept of a dynamical, expanding universe became the central pillar of cosmology.

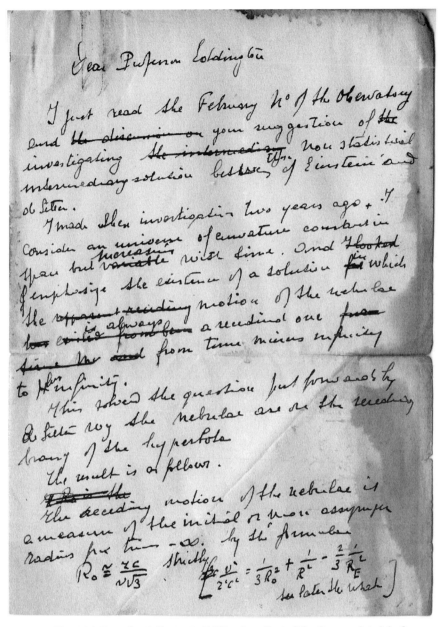

Fig. 11.1 Lemaître's letter to Eddington. First of the three undated draft pages, in which Lemaître tells Eddington that two years before he had found the solution that Eddington and de Sitter were looking for. (Archives Lemaître, Université Catholique, Louvain.)

Fig. 11.2 Sir Arthur S. Eddington (1882–1944). Eddington did much to disseminate the general theory of relativity.

Sir Arthur Stanley Eddington

Eddington was born in 1882 into a Quaker family, and was himself a lifelong member of the Society of Friends. His father was a headmaster in Kendal, south of the Lake District, but he died when the boy was two years old. His mother, with the son and his sister, four years his senior, moved to the south west of England. In 1901, he won a scholarship to Trinity College in Cambridge. With some interruptions he stayed in Cambridge until the end of his life in 1944. The longest spell away was his seven years as Chief Assistant at the Royal Greenwich Observatory from 1906 to 1913.

Contrary to an opinion often given about Eddington, he was very well acquainted with observational astronomy. In 1912, he went on an observing expedition to Malta to re-determine the longitude of a geodetic station, and in the same year he sailed with two assistants on his first solar-eclipse expedition to Brazil, where heavy rain prevented observations. In 1913, he was appointed Plumian Professor of Astronomy at Cambridge, and a year later Director of the Observatory, the posts he held until his death. He took up residence at the Observatory house in 1914, together with his mother and sister. Eddington observed the historic solar eclipse of 29 May 1919 from an island in the Gulf of Guinea, off the west coast of Africa. Its aim was to test a prediction of General Relativity about the bending of light. The weather was not ideal, but he could take photographs. With the results of another British expedition that had gone to Brazil, they confirmed Einstein's prediction (Dyson, Eddington and Davidson 1920). This confirmation was instrumental in spreading Einstein's fame among the general public.

Eddington very much liked outdoors activities, such as cycling and long, strenuous walking holidays. Of his personal inclinations his biographer says (Vibert Douglas 1956): 'Among the things forbidden by his strict upbringing were alcohol, tobacco and the theatre, but one by one he shed these early inhibitions, and in the course of time he patronised the theatre with discrimination, and he became a confirmed pipe smoker.' And, in his later years, he would not refuse a glass of wine.

The spreading of the gospel

By the middle of 1930, Eddington and de Sitter had widely advertised Lemaître's work, and incorporated its conclusion into their own researches. A generous recognition of Lemaître's discovery came from Eddington in 1930, when he wrote: 'Working in conjunction with Mr. G. C. McVittie, I began some months ago to examine whether Einstein's spherical universe is stable. Before our investigation was complete we learnt of a paper by Abbé G. Lemaître which gives a remarkably complete solution of the various questions connected with the Einstein and de Sitter cosmogonies. Although not expressly stated, it is at once apparent from his formulae that the Einstein world is unstable – an important fact which, I think, has not hitherto been appreciated in cosmogonical discussions.' He continues that in connection with the behaviour of spiral nebulae he had hoped to contribute some definitely new results, but this 'has been forestalled by Lemaître's brilliant solution'. In the same publication, Eddington showed that Einstein's universe is unstable (Eddington 1930). Together with Lemaître's findings and Hubble's observational confirmation this really meant the demise of the static universe.

Eddington was certainly sensitised to the possibility of a dynamic universe. In the Archives Lemaître at the UCL we find a letter, dated Louvain, 29 November 1925, in which Lemaître informs Eddington that he had been appointed lecturer at the University of Louvain, and thanks him for his support in this matter. At the end of the letter he says: 'I am sorry I could not yet send to you a copy of the "Note on de Sitter's universe" [this refers to Lemaître 1925b] that I showed you in July, but I did not receive them.' (Lemaître 1925c.) Thus Eddington had been personally informed of Lemaître's first struggle with de Sitter's intricate issue. Eddington was also aware of Robertson's fresh look at de Sitter (Robertson 1928); Eddington himself had communicated that paper to the *Philosophical Magazine*.

De Sitter also immediately adopted Lemaître in his publications. Already in May 1930, at the end of the article for which he incurred the wrath of Hubble – more about that later – de Sitter acknowledged Lemaître's non-static solution (de Sitter 1930b). And in June 1930 he followed it up with a lengthy discussion of Lemaître's solution (de Sitter 1930c), to which Lemaître replied in July in the same journal (Lemaître 1930b).

During 1930 and 1931, the pioneering work of Friedmann and Lemaître became generally known to those active in the field, in particular to Robertson and Tolman, about whose contributions we will have more to say later.

In 1931, Lemaître's publication of 1927 was even translated into English and published in the *Monthly Notices of the Royal Astronomical Society* (Lemaître 1931a). This was certainly an exceptional procedure and an extraordinary sign of appreciation, paying homage to a great discovery. However, some footnotes and a small, but historically significant section had been truncated. Lemaître's derivation of the 'Hubble constant' was omitted. It is not difficult to guess the reasons. In 1931, the search was still on for the right interpretation of the velocity–distance relationship, whereas its observational reality had been definitely established. With publication in *Monthly Notices*, Lemaître's theoretical work became finally available in English to all those who cared to know. But the observational material which convinced Lemaître of the validity of his theoretically derived Hubble law, and from which he had found the Hubble constant, had by the beginning of 1931 been superseded by newer data from Hubble, Humason and de Sitter. Thus, that part could be left out. – Still, why was it left out? – Nevertheless, it was an exceedingly exceptional procedure for *Monthly Notices* to re-publish a paper that had been published in another journal four years previously. It shows the importance that was attached to Lemaître's contribution.

Access to old issues of *Monthly Notices* is relatively easy, but because it is more difficult to find Lemaître's original article, we reproduce pages 55 and 56 in Fig. 11.3, which contain the truncated section in the 1931 English translation.

After Lemaître's publication in the March 1931 issue of the *Monthly Notices*, the expanding universe became the standard cosmological model, backed by the observational confirmation of Hubble, Humason and de Sitter. Lemaître had become an acknowledged authority in cosmology. He was invited to a British

— 55 —

période de la lumière reçue et δt, peut encore être considéré comme la période d'une lumière émise dans les mêmes conditions dans le voisinage de l'observateur. En effet, la période de la lumière émise dans des conditions physiques semblables doit être partout la même lorsqu'elle est exprimée en temps propre.

$$\frac{v}{c} = \frac{\delta t_o}{\delta t_i} - 1 = \frac{R_2}{R_1} - 1 \qquad (22)$$

mesure donc l'effet Doppler apparent dû à la variation du rayon de l'univers. *Il est égal à l'excès sur l'unité du rapport des rayons de l'univers à l'instant où la lumière est reçue et à l'instant où elle est émise.* v est la vitesse de l'observateur qui produirait le même effet. Lorsque la source est suffisamment proche nous pouvons écrire approximativement

$$\frac{v}{c} = \frac{R_2 - R_1}{R_1} = \frac{dR}{R} = \frac{R'}{R} dt = \frac{R'}{R} r$$

où r est la distance de la source. Nous avons donc

$$\frac{R'}{R} = \frac{v}{cr} \qquad (23)$$

Les vitesses radiales de 43 nébuleuses extra-galactiques sont données par Strömberg (¹).

La grandeur apparente m de ces nébuleuses se trouve dans le travail de Hubble. Il est possible d'en déduire leur distance, car Hubble a montré que les nébuleuses extra-galactiques sont de grandeurs absolues sensiblement égales (grandeur — 15,2 à 10 parsecs, les écarts individuels pouvant atteindre deux grandeurs en plus ou en moins), la distance r exprimée en parsecs est alors donnée par la formule log r = 0,2m + 4,04.

On trouve une distance de l'ordre de 10⁶ parsecs, variant de quelques dixièmes à 3,3 millions de parsecs. L'erreur probable résultant de la dispersion en grandeur absolue est d'ailleurs considérable. Pour une différence de grandeur absolue de deux grandeurs en plus ou en moins, la distance passe de 0,4 à 2,5 fois la distance calculée. De plus, l'erreur à craindre est proportionnelle à la distance. On peut admettre que pour une distance d'un million de parsecs, l'erreur résultant de la dispersion en grandeur est du même ordre que celle résultant de la dispersion en vitesse. En effet, une différence d'éclat d'une grandeur correspond à une vitesse propre de 300 Km. égale à la vitesse propre du soleil par rapport aux nébuleuses. On peut espérer éviter une erreur systématique en donnant aux observations un poids proportionnel à $\frac{1}{\sqrt{1+r^2}}$ où r est la distance en millions de parsecs.

(¹) Analysis of radial velocities of globular clusters and non galactic nebulae. *Ap. J.* Vol. 61, p. 353, 1925. M⁺ Wilson Contr. N° 292.

— 56 —

Utilisant les 42 nébuleuses figurant dans les listes de Hubble et de Strömberg (¹), et tenant compte de la vitesse propre du soleil (300 Km. dans la direction α = 315°, δ = 62°), on trouve une distance moyenne de 0,95 millions de parsecs et une vitesse radiale de 600 Km./sec, soit 625 Km./sec à 10⁶ parsecs (²).

Nous adopterons donc

$$\frac{R'}{R} = \frac{v}{rc} = \frac{625 \times 10^5}{10^6 \times 3,08 \times 10^{18} \times 3 \times 10^{10}} = 0,68 \times 10^{-27} \text{ cm}^{-1} \qquad (24)$$

Cette relation nous permet de calculer R₀. Nous avons en effet par (16)

$$\frac{R'}{R} = \frac{1}{R_0 \sqrt{3}} \sqrt{1 - 3y^2 + 2y^3} \qquad (25)$$

où nous avons posé

$$y = \frac{R_1}{R} \qquad (26)$$

D'autre part, d'après (18) et (26),

$$R_o^2 = R_a^2 y^2 \qquad (27)$$

et donc

$$3\left(\frac{R'}{R}\right)^2 R_a^2 = \frac{1 - 3y^2 + 2y^3}{y^3} \qquad (28)$$

Introduisant les valeurs numériques de $\frac{R'}{R}$ (24) et de R_a (19), il vient :

$$y = 0,0465.$$

On a alors :

R = R_a √y = 0,215   R_a = 1,83 × 10¹⁰ cm. = 6 × 10⁹ parsecs

R₀ = R y = R_a y^{3/2} = 8,5 × 10²⁸ cm. = 2,7 × 10⁸ parsecs

= 9 × 10⁸ années de lumière.

(¹) Il n'est pas tenu compte de N. G. C. 5194 qui est associé à N. G. C. 5195. L'introduction des nuées de Magellan serait sans influence sur le résultat.
(²) En ne donnant pas de poids aux observations, on trouverait 670 Km./sec à 1,16 × 10⁶ parsecs. 575 Km./sec à 10⁶ parsecs. Certains auteurs ont cherché à mettre en évidence la relation entre v et r et n'ont obtenu qu'une très faible corrélation entre ces deux grandeurs. L'erreur dans la détermination des distances individuelles est du même ordre de grandeur que l'intervalle que couvrent les observations et la vitesse propre des nébuleuses (en toute direction) est grande (300 Km./sec. d'après Strömberg), il semble donc que ces résultats négatifs ne sont ni pour ni contre l'interprétation relativistique de l'effet Doppler. Tout ce que l'imprécision des observations permet de faire est de supposer σ proportionnel à r et d'essayer d'éviter une erreur systématique dans la détermination du rapport σ/r. Cf. LUNDMARK. The determination of the curvature of space time in de Sitter's world M. N., vol. 84, p. 747, 1924. et STRÖMBERG, l. c.

Fig. 11.3 Lemaître's original derivation of the 'Hubble law' and 'Hubble constant'. Up to and including Equation (23), the main body of the text is the same as in the 1931 *Monthly Notices* (MN), though not the footnotes and references. The section after Equation (23), up to and including Equation (24) and the following line has in MN been replaced by: 'From a discussion of available data we adopt R'/R = 0.68 × 10⁻²⁷ cm⁻¹ and from (16)'. Then the translation comes back to the original text and continues with Equation (25); however, the footnotes on page 56 are not given (in MN the expression R'/R = 0.68 × 10⁻²⁷ cm⁻¹ carries Equation number 24). (Lemaître 1927, pages 55 and 56.)

Association Discussion on the 'Evolution of the Universe'. The illustrious participants included Sir James Jeans, W. de Sitter, Sir Arthur Eddington, Robert Millikan and E. A. Milne. The discussions are reported in *Nature*, 24 October 1931. Again, de Sitter pays tribute to Lemaître: 'Lemaître's theory not only gives a complete solution to the difficulties it was intended to solve, a solution of such simplicity as to make it appear self-evident, once it is known (like Columbus's famous solution of the problem of how to stand an egg on its small end) … There can be not the slightest doubt that Lemaître's theory is essentially true, and must be accepted as a very real and important step towards a better understanding of Nature.' (de Sitter 1931.) We come back to that meeting in Chapter 17, 'The seed for the Big Bang'.

Supreme consent arrived from Einstein. In April 1931, he published a short three-page report where he accepted the expanding universe (Einstein 1931b). We shall return to this publication in a later chapter.

But even before Einstein's approval, theoreticians had already turned to new questions: What causes the expansion? In the communication mentioned above, de Sitter suggested that the question was solved with the expanding universe. 'What becomes of the energy which is continually poured out into space?' His answer was that it was used up by the work done in the adiabatic expansion of the Universe. He also put his finger on a burning and unsolved problem. The age of the Universe, as deduced from the expansion velocity, is at most a few billion years, and this is short when matched against the times of stellar evolution – which at that time were grossly overestimated, and dangerously close to the age of our Earth.

Thus, the expanding universe had become an accepted fact. However, the model according to which it evolves had still to be found.

12

Hubble's anger about de Sitter

The publication of more accurate distances by Hubble, and of radial velocities by Humason in 1929, going far beyond those of Slipher, was a fascinating challenge to de Sitter. Such additional redshifts at large distances could provide a definite test for his theory; Lemaître's solution was still unknown to him at that time. Reading Hubble's 1929 announcement of a roughly linear velocity–distance relationship must have whetted his appetite for an even firmer statement. We have noted his report to the Royal Astronomical Society in January 1930 (de Sitter 1930a). In May 1930, his own study appeared in print (de Sitter 1930b). Here is Sandage's opinion about it: 'The paper was a model of Dutch thoroughness. Although the analysis was brilliant – it was far more complete than the one Hubble had provided in his discovery announcement a year earlier – the end result was the same as Hubble's: a linear relation.' (Sandage 2004, p. 504.) This paper was de Sitter's comeback to cosmology. During the 1920s, he had – when his directorship of the Observatory of Leiden allowed him time – worked on masses and longitudes of Jupiter's satellites.

De Sitter's check on Hubble

De Sitter first checked whether there was a direct relationship between apparent magnitudes and the logarithms of apparent (angular) diameters. He did that separately for spiral, elliptical and irregular nebulae. For distances, he extensively cited the 1926 publication of Hubble, and also that of 1929. The results gave him confidence that magnitudes could be employed for distance determinations. He then collected all the available observational data, including radial velocities for over fifty extragalactic nebulae. However, because of the accidental peculiar velocities, he excluded nebulae with radial velocities less than 300 km/s. He determined his Hubble constant from radial velocities up to 4000 km/s, thus four times the range that Hubble had considered in his diagram. De Sitter's result was 463 (km/s)/Mpc. He found that the safest way to find the Hubble constant – which, of course, at the time was not called by that name – was to draw a straight line in the velocity–distance diagram through the origin of the coordinate system and the centre of gravity of the nebulae considered. We remember that in 1927 Lemaître had decided to circumvent large individual uncertainties by relying on this method, and in 1929 Hubble had employed it as one amongst several methods.

De Sitter's result was basically the same as that of the two authors just mentioned. Then he suggested reversing the procedure; one could start with the measured radial velocities to derive distances. The linear velocity–distance relationship might give distances more reliably than nebular diameters or magnitudes, and he cited Hubble's contribution in the Mt Wilson Annual Report of 1928–29 (page 126) where the same suggestion had also been made.

Because he could now trust the linear velocity–distance relationship out to large distances, de Sitter attacks the problem that had probably been the primary reason for his investigation. What is the implication for cosmological models? The truly important result, he says, is that neither Einstein's solution, nor his own, can correspond to the truth. Then he refers to Lemaître's discovery of 1927, of which he had only become aware a few weeks previously. This *ingenious solution*, as he calls it, can provide the answer. He promises to return to a discussion of Lemaître in a separate communication.

Hubble's angry letter

Hubble was greatly angered by de Sitter's trespassing into 'his' velocity–distance domain without properly highlighting 'his' linear redshift–distance relationship (see Sandage 2004, p. 503). In the Huntington Library there is a letter dated 21 August 1930 from Hubble to de Sitter, complaining about de Sitter's publication of 1930 (Hubble 1930). We reproduce an excerpt here:

> The possibility of a velocity-distance relation among nebulae has been in the air for years – you, I believe, were the first to mention it. But our preliminary note in 1929 was the first presentation of the data where the scatter due to uncertainties in distance was small enough as compared to the range in distances, to establish the relation. In that note, moreover, we announced a program of observations for the purpose of testing the relation at greater distances – over the full range of the 100-inch, in fact. The work has been arduous but we feel repaid

since the results have steadily confirmed the earlier relation. For these reasons I consider the velocity-distance relation, its formulation, testing and confirmation, as a Mount Wilson contribution and I am deeply concerned in its recognition as such. In view of your casual reference "it has been remarked by several astronomers that there appears to be a relation …" I infer that you do not agree with my statement and therefore I would appreciate your reasons in order to reconsider the situation.

We are relieved that he did not challenge de Sitter to a duel with rapier, pistol or even heavier weaponry.

In the same letter, Hubble reproached de Sitter for having included in his analysis some of Hubble's published data:

Following the preliminary note we delayed our final discussion until sufficient new material was on hand to treat the subject in a comprehensive way. After two years work we are ready with some eighty velocities available, twenty-three of which are distributed in five clusters whose distances are determined, with a revision of the scale of distances, the scale of magnitudes, etc. Some of the new data has appeared in Annual Reports, etc., and most of them are known to other astronomers. We have always assumed that, where a preliminary result is published and a program is announced for testing the result in new regions, the first discussion of the new data is reserved as a matter of courtesy to those who do the actual work. Are we to infer that you do not subscribe to this ethics; that we must hoard our observations in secret? Surely there is a misunderstanding some where.

Well, the ethics of our time are that published data may be used by anyone in his or her own scientific research, as long as a proper reference is given to the origin of such data – and de Sitter did mention his sources.

Anyway, peace was eventually restored. There must have been a reply by de Sitter – which we could not locate, however – because on 23 September 1931, Hubble answered: 'Dear Professor de Sitter: Mr. Humason and I are both deeply sensible of your gracious appreciation of the papers on velocities and distances of nebulae. We use the term "apparent" velocities in order to emphasize the empirical features of the correlation. The interpretation, we feel, should be left to you and the very few others who are competent to discuss the matter with authority.' (Hubble 1931.)

Concerning Hubble's claim about formulation, testing and confirmation of the velocity–distance relationship, we can agree with testing and confirmation, but not with the claim to its first formulation. Hubble rediscovered it from observations alone, but the formulation and its central place in cosmology was first given by Lemaître on theoretical grounds, and he even provisionally tested it with observational data.

Also Robertson, inspired by theoretical considerations, suspected a linear velocity–distance relationship in the observational data: 'Comparing the data given by Hubble (1926) concerning the value l for the spiral nebulae with that of Slipher (see Eddington 1924) concerning the corresponding radial velocities, we arrive at a rough verification of $v \approx c \cdot l / R$.' (Robertson 1928.) Whether Hubble knew, or had heard about that prediction is an open question to which we will come back in Chapter 13, 'Robertson and Tolman join the game'.

When de Sitter published his investigation in early 1930, he referred to Hubble's work of 1929, but did not treat it as a great discovery. Instead, he said: 'It has been remarked by several astronomers that there appears to be a linear correlation between the radial velocities and the distances.' De Sitter could have been more generous. He might have mentioned that Hubble was the first to have collected observational data of sufficient quality to make a fairly convincing case for a linear velocity–distance relationship. We might add that Hubble himself was exceedingly partial in selecting his references (see also Sandage 2004, p. 504). Even in his influential *The Realm of the Nebulae*, published in 1936, he avoided any reference to Lemaître (Hubble 1936). Was he afraid that a gem might fall from his crown if people became aware of Lemaître's pioneering fusion of observation and theory two years before Hubble delivered the confirmation?

There is no reason to assume that de Sitter meant any offence towards Hubble. Throughout the 1920s, Slipher's redshifts and their relation to de Sitter's wavelength-shift predictions were the driving motor of the cosmological debate. Eddington, Wirtz, Silberstein, Lundmark and Strömberg participated in that debate. Then Lemaître and Robertson arrived, who combined a fresh look at the theory with new astronomical evidence. They still relied on Slipher's redshifts, but incorporated Hubble's distances of 1926. Both were satisfied that the observational data were consistent with their theoretically determined linear velocity–distance relationship. Thus, Hubble's strong evidence was certainly an important and welcome observational step, but not a revelation. It may be that Hubble was not sufficiently aware of what had been done in that field. When he wrote his letter to de Sitter, though, he ought to have known about Lemaître because de Sitter had mentioned him prominently.

The birth of the distance–redshift relation has a long history. In 1917, de Sitter threw a bone to the observers: if future observations confirm that spiral nebulae have systematically positive radial velocities, they would speak in favour of his cosmological theory – de Sitter knew of three nebular velocities at the time, one of which was negative. Wirtz, in 1922, found that the main motion of the nebulae could be described as a dispersal. In 1923, Eddington gave to his theoretically inclined readership Slipher's list of redshifts, with a hint that this might have something to do with cosmology. Wirtz and Lundmark thought that there might be a distance–velocity relation in 1924. Wirtz suspected a logarithmic increase, and Lundmark tried a parabolic fit, but the scatter in the data was too large to admit a definite statement. Also Strömberg in 1925 could not find sufficient evidence for the relation. With Hubble's further investigation of extragalactic nebulae in 1926, a new set of more reliable distances became available. In 1927, Lemaître found the linear velocity–distance relation from theoretical considerations. He matched it against new astronomical observations and found definite evidence for a positive correlation between distance and velocity, and plausible evidence for a linear relation. From these observational data he calculated the first Hubble constant. In 1928, independent of Lemaître, but based on about the same observations, Robertson saw in these data 'a rough verification' of the linear relationship. In 1929, Hubble, with improved and extended observational data of redshifts and distances, made an observational case for a linear relation, although he left the door open for a different law. One year later, de Sitter added further confidence to the linearity and, in 1931, Hubble and Humason confirmed the linear velocity–distance relation beyond doubt by significantly increasing distances and the sample size.

In this connection it may be appropriate to quote Eddington, one of the key figures in the cosmological debate of those years. In his book, *The Expanding Universe*, he wrote in 1933: 'The deliberate investigation of non-static solutions was carried out by A. Friedmann in 1922. His solutions were rediscovered in 1927 by Abbé G. Lemaître, who brilliantly developed the astronomical theory resulting therefrom. His work was published in a rather inaccessible journal, and it seems to have remained unknown until 1930 when attention was called to it by de Sitter and myself. In the meantime the solutions had been discovered for the third time by H. P. Robertson, and through him their interest was beginning to be realized. The astronomical application, stimulated by Hubble and Humason's observational work on the spiral nebulae, was also being rediscovered, but it had not been carried so far as in Lemaître's paper.' (Eddington 1933, p. 46.)

Just as there is no justification to glorify Hubble's publication of 1929 as the 'discovery of the expanding universe', and not even as the original discovery of the linear velocity–distance relationship, it would be unfair and wrong to belittle its importance. The publication was a significant milestone for cosmology. Hubble arrived at his velocity–distance relationship solely from observations. Unquestionably, this fact established the relation as a central criterion any theory had to fulfil.

Hubble and Humason enlarge the sample

Hubble and Humason continued to enlarge the sample to take in more distant objects, and together they verified the linear velocity-distance relation beyond doubt (Humason 1931, Hubble and Humason 1931). With justified pride they stated that the new data extended the distance to about

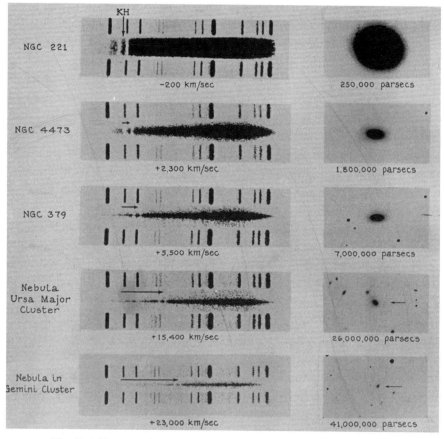

Fig. 12.1 Humason's spectra. A sample of Humason's spectra, published in 1936. The direct photographs show the linearly decreasing angular size of the first ranked cluster galaxy as the redshift increases. Practically the only recognisable features are the H and K lines of singly ionised calcium at 3968 and 3934 Å. The comparison emission spectra on each side of the nebular spectra are of helium. Calcium H and K lines are key lines in nebular spectra; in Chapter 4 we met them as prominent absorption features in the solar spectrum. (From Humason 1936.)

Table 12.1 *The Hubble constant H in (km/s)/Mpc determined up to 1931.*

Author	V_o	α	δ	H
Lemaître (1927) (1)	300	315°	+62°	625
Lemaître (1927) (2)	300	315°	+62°	575
Hubble (1929a) (3)	306	286°	+40°	465
Hubble (1929a) (4)	247	269°	+33°	513
Hubble (1929a) (5)	280	277°	+36°	500
Hubble (1929a) (6)	280	277°	+36°	530
de Sitter (1930b) (7)	286	314°	+66°	463
Hubble and Humason (1931) (8)				558

Lemaître: (1) when giving less weight to the most distant of his 42 objects, (2) giving equal weight to all observations. Hubble: (3) from 24 objects, (4) combining them into 9 groups according to direction and distance, (5) his favourite solution, (6) for 22 nebulae with unknown distances the mean distance is derived from the mean apparent magnitude, and then compared with the mean of their velocities. (7) De Sitter (1930b), from his Fig.10. (8) New velocities, mainly from Humason (1931). V_o, α, δ are the parameters of solar motion cited in the corresponding publications. The apex of solar motion is given by the listed α and δ coordinates.

eighteen times that available to Hubble in 1929. They found a Hubble constant of 560 (km/s)/Mpc.

Hubble and Humason left the interpretation to others. In their joint paper of 1931, they wrote: 'The present contribution concerns a correlation of empirical data of observation. The writers are constrained to describe the "apparent velocity-displacements" without venturing on the interpretation and its cosmological significance.' And Humason in his paper of 1931 clarified: 'It is not at all certain that the large redshifts observed in the spectra are to be interpreted as a Doppler effect, but for convenience they are expressed in terms of velocity and referred to as apparent velocities.' (Humason 1931.)

Hubble constants, extracted from publications, which, up to 1931, had explicitly dealt with the velocity–distance relation, are given in the table above. The largest velocity in Lemaître's list was 1800 km/s. In Hubble's plot it was 1100 km/s, but he had already access to velocity data up to 3800 km/s. De Sitter included in his plots points up to 8000 km/s. Hubble and Humason more than doubled that range and reached velocities close to 20 000 km/s. Qualitatively, the results of the different authors agree with each other.

In Chapter 9 on Lemaître, we mentioned that today we express redshifts less through conceptually misleading velocities, but rather with the parameter z, defined as $z = \Delta\lambda/\lambda$. From z we can calculate the change in the scale factor $R(t)$ from the value R_e, at the time when the light was emitted, to the value R_a, at the time of absorption by today's observer: $R_a/R_e = z + 1$. Thus, Lemaître's (1927) and Hubble's (1929a) top 'velocities' of between 1000 and 4000 km/s corresponded to $0.003 < z < 0.013$; Humason's (1936) extension to 42 000 km/s corresponded to $z = 0.14$. While on its journey, the light caught by Humason had seen the Universe increase in size by 14 per cent. The most distant galaxies seen in the Hubble Ultra Deep Field observations of 2003/4 lie at approximately $z = 6.5$. The Universe has grown by a factor of 7.5 since that light was emitted.

In the next chapter we shall report on Robertson's discovery in 1928 of the linear velocity–distance relation, $v = c \cdot (l/R)$. From that relation he derived R, which he called 'the radius of the observable universe'. His value for R would have corresponded to $H = 463$ (km/s)/Mpc. This agreed qualitatively with the other authors, which is to be expected because they all relied on practically the same observations.

13

Robertson and Tolman join the game

Robertson starts from first principles

In 1925, H. P. Robertson submitted his thesis on General Relativity to the California Institute of Technology. He afterwards spent two years in Germany. With his article 'On relativistic cosmology' he made his debut on the international cosmological stage. It dealt with de Sitter's empty world. It was written in Göttingen and communicated to the *Philosophical Magazine* by Eddington. Its purpose was to replace de Sitter's line element by 'a mathematically equivalent solution which is susceptible of a perhaps simpler interpretation and in which many of the apparent paradoxes inherent in [de Sitter's solution] were eliminated.' (Robertson 1928.)

Robertson did not seem to have known about the previous investigations of Friedmann and Lemaître. In his paper he cites Einstein, de Sitter, Eddington, Weyl, Silberstein, Lundmark and Hubble. Like Lemaître one year before, he found a representation that respects homogeneity of space. But unlike Lemaître, Robertson clung to de Sitter's stationary universe and his 'dynamical' interpretation. He saw that his line element was dynamical; it depended on time. But Robertson looked for a static solution, which he wanted to achieve by transformations of the coordinate system.

He also arrived at the formula that in Lemaître's hand had become the velocity–distance relation. Restricting himself to cosmologically not very distant objects, but sufficiently far removed so as not to be disturbed by local effects, he found a linear correlation between 'assigned velocity' v, the distance l to the nebula, and R, which he called the 'radius of the observable world'. He wrote this as $v = c \times (l/R)$. Occasionally it is claimed that Robertson derived and calculated the Hubble constant in 1928. This is a misinterpretation. Whereas

Lemaître had employed this same formula to write down the velocity–distance relation – later called the Hubble law – Robertson derived the size of R from the observed v and l. To obtain a numerical value, he compared velocities from Slipher with distances Hubble had published in 1926, and arrived at what he considered to be a 'rough verification' of that linear relation. For the radius of the Universe, he found $R = 2 \times 10^{27}$ cm. His c/R corresponds to the Hubble constant. With $R = 2 \times 10^{27}$ cm, $H = 463$ (km/s)/Mpc, which is close to the values found by Lemaître one year before, and Hubble one year later, but Robertson did not calculate such a number.

Although Robertson did not calculate the Hubble constant, he did write down the linear velocity–distance relation. As we have noted before, according to Sandage, Hubble never hinted that he knew about Robertson's linear velocity–distance relation, whereas Robertson told Sandage he had discussed his prediction with Hubble in 1928 (Sandage 2004, p. 501).

In 1929, Robertson returned to the cosmological problem (Robertson 1929). From the very beginning he neatly separated spacetime into space and time: $ds^2 = dt^2 + g_{ij}\, dx^i\, dx^j$, where the spatial metric tensor, g_{ij}, is a function of time, t: $g_{ij}(t)$. He had in the meantime become aware of Friedmann's work and Lemaître's first investigation of 1925. Yet, Robertson remained within a stationary world. He had gone a step beyond de Sitter's mysterious slowing down of time, but this change of approach did not lead to a new concept of the Universe.

Solutions of Einstein's equations have in general no special symmetries. However, to describe the large-scale structure of a spatially homogeneous universe, it is usual to separate 4-dimensional spacetime into a spatial volume and a time direction. In addition, we treasure the Copernican principle of 'no special place for us' and therefore generalise observations to the hypothesis of universal homogeneity and isotropy. These premises imply symmetries in the solutions of Einstein's equations. Robertson was the first to explicitly look for all the mathematical universes that would satisfy these physical requirements.

For the formal development of cosmology, Robertson's publication of 1929 was an important step forward. He enumerated the properties demanded for cosmological models. The description of spacetime should be separated in three spatial coordinates and one time coordinate. He assigns x^0 to time. The matter of the Universe has time-like geodesics, $x^i = $ constant as world-lines, where spatial coordinates take the indices $i = 1$, 2, 3. The coordinate, t, can then be interpreted as a mean time, which serves to define proper time and simultaneity for the whole universe. The 3-dimensional x-space will be interpreted as the 3-dimensional physical space at time, t. To express the intrinsic uniformity of the Universe, all spatial directions are fully equivalent in the sense that one cannot distinguish between them by any intrinsic property of spacetime. And there should be no difference between the observations of any two contemporary observers. Robertson then repeats these conditions in more mathematical language.

Robertson points out that his prescriptions express in another way Weyl's assumption (Weyl 1923b) that the world-lines of all matter in the Universe form a coherent pencil of geodesics.

Robertson also turned to the Doppler effect. The material content of his idealised universe is at rest, thus no classical Doppler shift is to be expected. However, the relation between his time and length scales vary according to a time-dependent function, $f(t)$. This function describes the relation between the frequency of an emitted spectral line and the frequency of the same line when the photon arrives at the observer. For the time interval that corresponds to the inverse of the frequency, he wrote $\Delta t = \Delta t_0 \cdot \exp(f(t) - f(t_0))$, where t_0 refers to the time of emission and t to the time of absorption. For the redshift, he derived $\Delta\lambda/\lambda = \Delta t/\Delta t_0 - 1$, which he could then express through $f(t)$. However, Robertson only briefly commented on this result. He remarked that of the two stationary universes, only that of de Sitter will show such an effect, because in Einstein's universe $f(t)$ is a constant; but he did not describe the form of $f(t)$, nor its significance. He concluded: 'The material content of this idealized background of the actual universe, being at rest with respect to the special coordinates, leads to a unique Doppler shift which will appear in the actual universe as a residual effect.'

When reading Robertson's publications of 1928 and 1929, one is held in permanent suspense. He develops the proper formalism for a dynamic universe. At any moment, his jump into an expanding universe is expected. After all, in his publication of 1928 he shows himself familiar with all the observational evidence that in 1927 had served Lemaître in his discovery. Yet, whenever we expect the jump, Robertson branches off to show how the static universe can be accommodated in his formalism, or he simply stops short of a physical interpretation. In particular, in his publication of 1929, where he very neatly develops a dynamical solution of cosmology and derives the redshift from a time-dependent factor, he does not discuss the cosmological significance in his conclusion.

Tolman and the annihilation of matter

With a burst of publications in 1929, Richard Tolman joined the theoretical debate about wavelength shifts in extragalactic nebulae. In his two initial publications, which appeared practically simultaneously, he remained firmly within de Sitter's frame, but shifted the emphasis from a static to a 'steady' universe. He derived an expression for radial acceleration, d^2r/dt^2, and found that nebulae should disperse away, thus emptying space around the observer. Tolman acknowledged that de Sitter, Weyl, Eddington and Silberstein had already pondered about implications of such an acceleration. He was puzzled by the apparent contradiction between the theoretical tendency to scatter and the observed uniform nebular distribution. For him it turned into a choice between two alternative pictures. They may be descriptively labelled 'hypothesis of continuous entry' and 'hypothesis of continuous formation'. In his ensuing discussion, Tolman put much weight on the crossing of perihelion by nebulae, and thus appearing or disappearing in and out of the field of observation (Tolman 1929a,b).

When reflecting on the qualitative difference between the Einstein and de Sitter universe, he pondered that by lowering the actual amount of matter in our Universe, it would approach the form demanded by de Sitter, and the dimension would be determined by the cosmological constant, Λ. If, however, Einstein's model were correct, then by gradually decreasing the amount of matter, the Universe would get smaller due to the relation between the total mass and Λ, and finally, in the complete absence of matter, the Universe would shrink to zero volume (Tolman 1929b, p. 247).

Tolman was not satisfied with his results and in 1930 brought a new twist to the tale: the annihilation of matter. He had become aware of the dynamic line element of Robertson, and agreed that no static line element could successfully account for the redshift. But what could justify a non-static line element? Well, he said, 'the actual line element may well be non-static if the matter in the universe is not in a steady state.' He then introduced into the discussion his new flash of inspiration: 'As the basic hypothesis … we shall assume … that the matter of the universe is not in a steady state but rather that a general transformation of matter into radiant energy is taking place throughout the universe at the rate necessary to account for the radiation from stellar objects. If such a process is going on, the line element for the universe cannot be static, but necessarily must be non-static, since matter and the radiation produced from it would not have the same effect on the gravitational field.' (Tolman 1930a.)

He proceeded 'to deduce the form of line element which would correspond to the transformation of matter into radiation'. In this line element he kept the 'radius of the universe' at a fixed size, R. In his conclusions, he pointed to similarities between his line element and that of Einstein. The distinction lies in a coefficient that measures the average rate at which matter is changing into radiation. If the coefficient were zero, his line element would reduce to that of Einstein's. He therefore concluded that 'we can describe the model of the universe which corresponds to our line element as a non-static Einstein universe – necessarily non-static since it takes into account a general process of transformation of matter into radiation ...' (Tolman 1930a.) In his follow-up publication he further elaborated on that model (Tolman 1930b).

Tolman then became aware of Friedmann and Lemaître through contacts with Robertson, Eddington and de Sitter. This led to an additional publication, where he compared the different approaches. In addition, he incorporated Lemaître's expanding universe into his approach. He conceded that a non-static line element could be obtained from an expanding universe without introducing any annihilation of matter. In fact, he sees no conflict between the two qualitatively quite different mechanisms: annihilation itself may, under appropriate conditions, lead to an expansion of the Universe. To support his claim he derived a relation between the mass lost in annihilation, the pressure, the density and the metric tensor, g. He then showed that under certain conditions, annihilation of matter leads indeed to an expanding universe (Tolman 1930c).

For a short spell, conversion of matter into radiation also captured the attention of de Sitter. In the wake of Lemaître's non-static solution he pondered about the many possibilities of constructing world models. When he became aware of Tolman's suggestion that annihilation was a means of influencing the line element and explaining observed redshifts, he packed it into a 'new theory': 'The new theory thus incidentally gives an answer to the old question what becomes of the energy that is continually being poured out into space by the stars. It is used up, and more than used up, by the work done in expanding the universe. Nevertheless it would not be correct to say that the universe is expanded by the radiation pressure. It would expand just the same if γ [rate of conversion of matter into radiation] were zero, i.e. if no radiation was emitted by matter. The expansion is due to the constant λ [Einstein's cosmological constant].' (de Sitter 1930c.) Annihilation does occur, but it is not the decisive factor for the expansion of the Universe. That was de Sitter's opinion. He left that blind alley after having marked the territory with some of his own scent, and did not return.

Tolman would not abandon annihilation. However, he did not insist on its being the ultimate explanation for the observed redshifts. In his voluminous *Relativity, Thermodynamics and Cosmology*, published in 1934, he inserted a small section on 'The relation between red-shift and rate of disappearance of matter', where he pointed to observational problems: reasonable assumptions on the conversion rate would not lead to a linear redshift–distance relation (Tolman 1934).

Tolman's book contains a personal evaluation of what in 1934 could safely be stated about the Universe, and what was still unknown. We summarise the conclusions of his last chapter (Tolman 1934):

> The data are insufficient to provide a precise cosmological model,
> suitable for treating events far distant from our own location and time.

It seems reasonable to conclude that the expansion of the universe is a phenomenon which has progressed during at least a hundred million years and which will presumably continue for a comparable time.

We have no knowledge as to conditions in the actual universe beyond some 10^8 light years. Hence it is entirely possible that other densities of distribution, or contraction instead of expansion, may be present in portions of the universe beyond the reach of our present telescopes. We must be careful not to substitute the comfortable certainties of some simple mathematical model in place of the great complexities of the actual universe. When adopting a homogeneous model as a satisfactory first approximation, we do not have sufficient information to distinguish definitely between a closed, an open spatially curved, or an open spatially uncurved universe.

Weyl's brief return to cosmology

In the middle of 1929, Hermann Weyl spent some time in Berkeley. At a conference in Princeton he had met Robertson and they discussed their respective models (see Weyl 1923c and Robertson 1928). They agreed that their cosmology was identical. Weyl wanted to put it into print and at the same time answer some critical questions of Tolman. Weyl had in the meantime become acquainted with Lemaître's first publication of 1925, but he was, as yet, unaware of Friedmann (1922, 1924) and Lemaître's fundamental discovery of 1927. He began his paper, 'Redshift and relativistic cosmology', by pointing to the observational progress in astronomy (Weyl 1930):

Recent observations made on spiral nebulae have ascertained their extragalactic nature and confirmed the redshift of their spectral lines as systematic and increasing with the distance (Hubble 1929). By these facts the cosmological questions about the structure of the world as a whole, to which general relativity gave rise in a purely speculative form, have acquired an augmented and empirical interest. It is not my opinion that we can vouch for the correctness of the "geometrical" explanation which relativistic cosmology offers for this strange phenomenon with any amount of certainty at this time. Perhaps it will have to be interpreted in a more physical manner, in correspondence with the ideas of F. Zwicky [Zwicky 1929]. But the cosmologic-geometrical conception must on any account be examined seriously as a possibility.

Zwicky had tried to explain the observed redshifts in terms of a kind of 'tired photon'. Weyl then referred to the line element given by Robertson, where time and space are split, and he acknowledged that Lemaître had already given the same representation in 1925. Weyl wrote the line element qualitatively in the same form. To the constant that formerly represented the radius of space, he gave a new meaning: 'its simplest interpretation is that of being the standard of measurement for the scattering of the stars or the redshift which corresponds to it as Doppler effect.' (Weyl 1930.)

'Scattering of stars' is Weyl's new suggestion to explain redshifts. It corresponds to the classical Doppler effect stretched to cosmic dimensions and presented in cosmological terms. It is a dynamical universe. But it places dynamics within space and does not advocate the dynamics of space. Robertson (1933) qualified Weyl's publication as 'Restatement of 1923 in coordinates of Lemaître (1925) and Robertson (1928)'.

For Weyl, the de Sitter hyperboloid presents the homogeneous state of the world. He asks about the geometry that corresponds to the real world. Friedmann had also discussed these questions earlier (Friedmann 1922, 1924) but, from the lack of citations, we have to assume that Weyl was not yet aware of this pioneer of cosmology, nor of Lemaître's publication of 1927.

Whereas in his earlier investigation Weyl saw de Sitter's empty world as a universe where matter had been shifted to the horizon, he no longer had to deal with that problem. Instead, he advocated a universe where stars have been in mutual interaction since eternity. He rejects Tolman's ideas of stars appearing and disappearing across the horizon. Although Weyl's paper contains no new perspectives, it shows that progress in astronomical observations had incited the theoreticians to return to cosmology. This was largely due to the 100-inch Hooker telescope on Mount Wilson, of which Hubble had made very good use.

14

The Einstein–de Sitter universe

Einstein had discarded Friedmann as well as Lemaître. Now that luminaries like Eddington and de Sitter had adopted Lemaître's expanding universe, would Einstein change his opinion? Yes, he did. Why?

Einstein's conversion

Eddington's biographer tells us that Einstein stayed with Eddington and his sister in June 1930 (Vibert Douglas 1956, p. 102). In the biography we even see a photograph of Einstein and Eddington in the latter's garden, taken by his sister. At that time, Eddington was very much involved in cosmology. He had enthusiastically adopted Lemaître's expanding universe, and his article 'On the Instability of Einstein's Spherical World' had just appeared in *Monthly Notices*. The conversation of the two friends must doubtless have turned to the central issue of cosmology. We may therefore take it for granted that Einstein learnt, at the latest in June 1930, about de Sitter and Eddington's adoption of Lemaître's expanding universe, and about the supporting observational evidence from Hubble and de Sitter (Hubble 1929a, de Sitter 1930b).

It is occasionally recounted that Einstein was converted to the expanding universe when Hubble showed him observational evidence during his visit to the Mount Wilson Observatory at the end of of that year. This story is very unlikely. We have Einstein's diary of his visit (Einstein 1930). He and his second wife arrived in San Diego early in the morning on 31 December. They had breakfast on board the ship with the mayor of San Diego and were welcomed by boys and girls with flowers. Interviews with journalists followed, a radio interview, then another reception in town with organ music and speeches, and finally a three-hour car ride along the seashore to Pasadena in the afternoon.

The pages of the diary are amusing to read: 'Gestern Abend bei Millikan, der hier die Rolle des lieben Gottes spielt.' (Last night at Millikan's home, who plays here the part of our Good Lord.) But let us turn to cosmology. The entry on 3 January 1931 reads: 'Arbeiten am Institut. Zweifel an Richtigkeit von Tolmans Arbeit über kosmologisches Problem. Tolman hat aber Recht behalten.' (Work at the Institute. Doubt about correctness of Tolman's work on the cosmological problem. But Tolman was right.) What a pity Einstein does not tell us where he thought Tolman was right. Then there are many entries about contacts with the solar group and St John, a member. This is not astonishing. Sandage tells about the efforts of St John to find Einstein's gravitational redshift in the solar spectrum (Sandage 2004, p. 135). Changing the subject: 'Heute erhielt ich eine wundervolle Guarneri-Geige geliehen. Gibt eine Bomben-Reklame für den Händler.' (Today I received on loan a wonderful Guarneri violin. This will result in a roaring advertisement for the dealer.) Then he enumerates some of his activities between 20 and 25 January: 'Radio interviews, scientifically interest-ing things, astronomical plates with Doppler-shifts of nebulae, talk with St John about redshift in the Sun, magnetic field in sunspots … lecture of the excellent Tolman about relativistic thermodynamics, overcame a slight influenza.' (Translated from German.) He returned to Europe in March 1931.

But where is mention of his contact with Hubble? He certainly did meet him. They can be seen together on photographs, some of them taken on Mt Wilson, for example, Einstein and Hubble at the telescope or on 31 January 1931, in a group of five at the solar tower (Sandage 2004, p. 142). If Hubble and Einstein had extended discussions, they did not impress Einstein sufficiently to be entered in his diary. There is the brief remark, 'astronomical plates with Doppler-shifts of nebulae'. That hardly sounds like a great revelation, but rather like a tourist's curiosity, without any further comment. In any case, a glance at the spectra contained little scientific information. The devil was hiding in the intricate distance determination to the nebulae. On the other hand, his diary shows he was much impressed by Tolman, and meeting him was probably more influential on his cosmological conversion than any meeting with Hubble.

Ms Barbara Wolff at the Albert Einstein Archives in Jerusalem suggested another reason for the apparent lack of close contact between Einstein and Hubble (personal communication, 2007). At that time, Einstein was not very fluent in English, and Hubble was probably not very fluent in German. Tolman, however, acted on several occasions as Einstein's more or less official translator. Thus Einstein might well have preferred to obtain his information, including those concerning observations, from Tolman rather than from Hubble. In addi-tion, Hubble and Einstein would hardly have talked about theory; this was not Hubble's domain. A photograph in the Library of Caltech (Photo ID 1.8-5) may substantiate these presumptions. It is a portrait of Einstein and carries a dedica-tion to Tolman in the form of a little poem composed by Einstein, which, translated into English, begins 'Many things connect us, not only relativity', testifying to the friendly relations between them.

Einstein certainly did not underestimate the importance of Hubble's observa-tions. They provided unequivocal evidence for a non-static universe. His appre-ciation is born out by a letter to his close friend Besso, dated 1 March 1931. There is no indication from where it was sent but, if it is correctly dated, it must have been written on Einstein's journey to the east coast of America before taking the boat back to Europe. He mentioned that his trip to America was very interesting but strenuous. Then he says: 'Die Leute vom Mount Wilson-Observatorium sind ausgezeichnet. Sie haben in letzter Zeit gefunden, dass die Spiralnebel räumlich annähernd gleichmässig verteilt sind und einen ihrer Distanz proportionalen wüchsigen Dopplereffekt zeigen, der sich übrigens aus der Relativitätstheorie zwangslos folgern lässt (ohne kosmologisches Glied). Der Hacken ist aber, dass die Expansion der Materie auf einen zeitlichen Anfang schliessen lässt, der 10^{10}, bzw. 10^{11} Jahre zurückliegt. Da eine anderweitige Erklärung des Effektes auf grosse Schwierigkeiten stösst, ist die Situation sehr aufregend.' (Einstein 1931a.) (The people of the Mount Wilson Observatory are excellent. They have lately found that the spiral nebulae are spatially nearly evenly distributed and that they show a Doppler shift proportionally growing with their distance, which, by the way, follows without problem from the theory of relativity (without the cosmological term). However, the snag is that the expansion of matter points to a beginning in time that lies 10^{10}, respectively 10^{11} years in the past. As any other explanation meets with great difficulties, the situation is very exciting.)

The cosmological question had obviously recaptured his attention. He talks about 'the people from Mount Wilson'. He does not single out any individual. We saw that he was highly impressed by Tolman, who was actually affiliated to Caltech, and working intensively on his cosmological model. Einstein had also been shown the observational work on galaxies. In April 1931, Hubble and Humason would submit their 38-page landmark paper on the velocity–distance relation. Einstein had certainly been informed about these activities. And he was, no doubt, aware that Hubble was the leading force behind the cosmological observations. Thus, his extended discussions with Eddington and Tolman, as well as personal contact with Hubble and his observational evidence must have been fermenting at the back of his mind during the return journey to Europe.

Back in Germany, we know what Einstein did on the evening of 13 April from his diary: 'Nach dem Abendessen Haydn-Trio. Abends im Studierzimmer inter-essante Idee zum kosmologischen Problem … Abhandlung zum kosmolo-gischen Problem begonnen.' (After dinner Haydn-Trio. In the evening in my study interesting idea about the cosmological problem … Began the essay on the cosmological problem.) And on 16 April: 'Kosmolog. Arbeit eingereicht.' (Cosmological work submitted.) Indeed, in April 1931 there appeared a short, three-page report in the proceedings of the Königlich Preußischen Akademie der Wissenschaften: Einstein accepted the concept of an expanding universe (Einstein 1931b).

He began the report by mentioning Hubble's observations of redshifts; they invalidate his own initial assumption of a static universe. Then he says: 'Es ist von verschiedenen Forschern versucht worden, den neuen Tatsachen durch einen sphärischen Raum gerecht zu werden, dessen Radius P zeitlich veränderlich ist. Als Erster und unbeeinflusst durch Beobachtungstatsachen hat A. Friedman diesen Weg eingeschlagen, auf dessen rechnerische Resultate ich die folgenden Bemerkungen stütze.' (Several investigators have tried to cope with the new facts by using a spherical space whose radius, P, is variable in time. The first who, uninfluenced by observations, tried this way was A. Friedmann, on whose mathematical results the following remarks will be based.) He then recalls how his own static solution of 1917 looks in the notation of Friedmann, and points out that it is not stable. In the course of time, it will diverge increasingly from the stable state. 'Schon aus diesem Grunde bin ich nicht mehr geneigt, meiner damaligen Lösung eine physikalische Bedeutung zuzuschreiben, schon abgesehen von Hubbels [sic!] Beobachtungsresultaten.' (Already on this account I am no longer inclined to ascribe a physical significance to my former solution, quite apart from Hubbel's [sic!] observational results.) And he continues: 'Unter diesen Umständen muss man sich die Frage vorlegen, ob man den Tatsachen ohne die Einführung des theoretisch ohnedies unbefriedigenden λ-Gliedes gerecht werden kann.' (Under these circumstances one has to ask the question, whether one could not do justice to the facts without introducing the theoretically anyway unsatisfactory λ-term.)

Einstein then adopted Friedmann's oscillating solution. Assuming a mean density of the Universe of $\rho \approx 10^{-26}\,\mathrm{g/cm^3}$, he derived an oscillation period of approximately 10^{10} years, a number that had already been given by Friedmann in 1922. He was fully aware that this period was short compared with the then accepted ages of the Sun and the Earth. But he was not unduly worried; he thought the difficulty could be avoided by invoking inhomogeneities in the distribution of stellar matter, which invalidate the approximations of the theoretical homogeneous treatment. Anyway, Einstein remained optimistic. In his opinion, the simplicity of the theory favoured a comparison with astronomical facts; it also showed how careful one had to be in astronomy when extrapolating over long periods. He ended by emphasising that General Relativity could account for the observational facts without the cosmological term.

Einstein's note of 1931 is astonishingly short on references. Friedmann (1922) is the only publication he cites. His remark that 'Several investigators have tried to cope with the new facts' is, to put it mildly, a rather gross understatement. Lemaître had given a solution in 1927 that coped very well with the new facts. Eddington had shown that Einstein's solution was unstable in 1930; Robertson, Tolman and de Sitter had dealt with the new situation as well. Einstein blithely ignores them. However, we should not be too harsh on him. He did not claim a new discovery. His main purpose was to get rid of the cosmological constant.

Einstein's letter to Tolman of 27 June 1931 shows again what a relief it must have been for him to be rid of the cosmological constant. Einstein wrote in German, Tolman answered in English. The main point of his publication, he said, was to point out that Λ became unnecessary if one admitted a time-variable radius of the Universe (Weltradius). 'Dies ist ja wirklich unvergleichlich befriedigender.' (Einstein 1931c.) (That is indeed incomparably more satisfactory.) Just as in his published note and in his letter to Besso, Einstein did not conceal his concerns about the beginning of the Universe. However, he said, they can be resolved by pointing to the inhomogeneities in the density, because they invalidate the calculations based on homogeneity. He thinks that in the very early Universe, space must have been relatively inhomogeneous. In this connection, he refers to a suggestion of Lindemann that such inhomogeneities might have been instrumental in the formation of planets. Einstein ends his letter with a postscript: 'Wie Sie sehen, bin ich wieder davon abgekommen, Hubbels [sic!] Linienverschiebungen in irgenwie abenteuerlicher Weise zu deuten.' (As you can see, I abandoned my attempts to interpret Hubbel's [sic!] line shifts in some adventurous fashion.)

Tolman answered on 14 September 1931. He thought Lindemann's suggestion interesting and he liked Einstein's choice of a quasi-periodic universe. But he was less happy about the dismissal of Λ, although 'it becomes no longer necessary to inquire into the significance and magnitude of what would otherwise be a new constant of nature. On the other hand, since the introduction of the λ-term provides the most general possible expression of the second order which would have the right properties for the energy–momentum tensor, a definite assignment of the value $\lambda = 0$, in the absence of an experimental determination of its magnitude, seems arbitrary and not necessarily correct.' (Tolman 1931.) Tolman ends the letter with a postscript: 'You will be interested to know that we had another sitting with Upton Sinclair's medium and again nothing happened.'

Frederick A. Lindemann (1886–1957) did his Ph.D. with Nernst in Berlin. He worked on specific heats and later became closely associated with Churchill's war efforts. He collected Einstein from Southampton on 1 May 1931 and took him to Oxford. That evening Einstein had to dress in a dinner jacket for a dinner with 500 professors and students, all male, even the waiters. 'Man kriegt eine leise Idee, wie grässlich das Leben ohne Weibsleute wäre.' (Einstein 1930.) (One gets a faint idea how awful life would be without women.) During the following days, Einstein and Lindemann talked about cosmology on several occasions.

Another aspect of Einstein's 1931 publication attracts attention. He stresses the fact that Friedmann found the dynamic universe uninfluenced by observations. He may well have kicked himself for not having discovered the dynamical solution on purely theoretical grounds in the first place.

The publication did not add any new insights to cosmology. Its historical interest lies in Einstein's official rejection of what he now calls the 'theoretically unsatisfactory' cosmological constant. Why had he changed his mind? Einstein accepted the dynamic universe only when he realised that his 'static' solution was unstable, and therefore did not represent a truly static universe. Already in 1922, and then again in 1927, he had been confronted with dynamic solutions, first by Friedmann and then, even personally, by Lemaître. Twice he refused to see in these alternatives anything but mathematical vagaries. It must have been Eddington's 'On the instability of Einstein's spherical world' which showed him that his own solution did not deliver what he had hoped for. Discussions with Tolman acquainted him with the work done at Mount Wilson. Hubble's observations lent additional credibility to the non-static, expanding universe. On their own they would hardly have meant much to Einstein. After all, Lemaître had already confronted him with the observational evidence in 1927.

We have mentioned twice that Einstein misspelt Hubble's name as 'Hubbel': in the letter to Tolman, and in his publication of 1931, where he mentions Hubble's name five times but gives no reference to his publication. This begs the question whether Einstein had actually read Hubble's publication, or whether he was satisfied with the information he had received personally from Hubble, but probably mainly from Tolman.

Einstein's banishment of the cosmological constant deterred neither Eddington nor Lemaître. They had been debating about the origin of the expanding universe, and Λ was playing a crucial part; they were certainly not going to abandon that gem, excavated and then discarded by Einstein.

Einstein and de Sitter agree on the structure of the Universe

At the beginning of December 1931, Einstein left Germany for another trip to California. The sea was rough, it was a stormy crossing, and the political current in Germany gave him big worries. On 6 December, he wrote in his diary: 'Heute entschloss ich mich, meine Berliner Stellung im Wesentlichen aufzugeben.' (Today I decided to essentially give up my position in Berlin.) He arrived in Los Angeles on the evening of 30 December. Next morning, Tolman came to the ship to collect Einstein and his wife (Einstein 1932).

Einstein was very busy with social engagements of all kinds, and also with several scientific projects. On 8 January 1932, he scribbled in his diary: 'Abfassung einer Notiz mit De Sitter über Beziehung zwischen Hubbel-Effekt [sic!] und Materie-Dichte der Welt bei Vernachlässigung der Krümmung und des λ-Gliedes.' (Einstein 1932.) (Formulation of a draft with de Sitter about the relation between the Hubbel-effect and matter density of the world, neglecting curvature and the λ-term.) A photograph exists of that time, where Einstein and de Sitter are discussing at the blackboard. This draft would become the Einstein–de Sitter universe.

Together they produced a two-page note 'On the Relation between the Expansion and the Mean Density of the Universe'. If it had not been for the two famous names, the paper would hardly have passed any refereeing system. What it said had all been said before by Friedmann, Lemaître and Robertson. However, here were the authors of the first two rival cosmological models agreeing about the structure of the Universe. That was certainly worth publishing, and they only needed two pages for their message (Einstein and de Sitter 1932). The publication was communicated by the Mount Wilson Observatories on 25 January 1932, and appeared in the 15 March issue of the *Proceedings of the National Academy of Sciences*.

We saw that Einstein had made his peace with the expanding universe early in 1931, and de Sitter had been an eager advocate since his first encounter with Lemaître's work early in 1930. At the beginning of their joint paper they refer to a publication by Heckmann (1932), who 'pointed out that the non-static solutions of the field equations of the general theory of relativity with constant density do not necessarily imply a positive curvature of three-dimensional space, but that this curvature may also be negative or zero.' Indeed, already in 1931 and then again in 1932, Heckmann had emphasised that in addition to the spherically closed and the hyperbolically open space solutions of the field equations, there also existed solutions in a spatially infinite, flat Euclidean space. However, neither Heckmann, nor Einstein, nor de Sitter mention that this possibility had already been included in the cases looked at by Friedmann and Robertson.

Einstein and de Sitter were probably looking for the simplest formal solution giving an expanding, homogeneous, isotropic universe that produced the observed redshifts. 'There is no direct observational evidence for the curvature, the only directly observed data being the mean density and the expansion, which latter proves that the actual universe corresponds to the non-statical

Fig. 14.1 Einstein and de Sitter. Photographed 8 January 1932 at California Institute of Technology, Pasadena. (Leiden Observatory Archives.)

case.' And, concerning the cosmological term Λ (which they called λ) they write: 'Historically the term containing the "cosmological constant" λ was introduced into the field equations in order to enable us to account theoretically for the existence of a finite mean density in a static universe. It now appears that in the dynamical case this end can be reached without the introduction of λ.' It is well worth paying attention to the wording. They did not treat the case $\Lambda = 0$, they went further and excluded Λ from their formalism.

No spatial curvature, and the cosmological constant banished, implies a definite relation between the rate of expansion, expressed in the Hubble constant, H, and the mean 'critical density', ρ_c. With $H = 500\,(\text{km/s})/\text{Mpc}$ they obtained $\rho_c = 4 \times 10^{-28}\,\text{g/cm}^3$. This is the upper limit of that found by de Sitter from astronomical observations.

Looking at the mean density, ρ, of their model, they admit that it might be on the high side. However, 'it certainly is of the correct order of magnitude, and we must conclude that at present time it is possible to represent the facts without assuming a curvature of three-dimensional space. The curvature is, however, essentially determinable, and an increase in the precision of the data derived from observations will enable us in the future to fix its sign and to determine its value.' Thus, they are not dogmatic about 'no curvature'.

In the Einstein–de Sitter universe, no cosmological term, Λ, is active. The density of matter is chosen such that the contracting effect of gravitation decelerates the expansion velocity asymptotically to zero; but it takes an infinite time to reach that point. The result is an open, spatially flat Euclidean universe. It should be stressed that although space is flat, spacetime is not flat; the curvature of spacetime is determined by the deceleration of the expansion velocity.

The Einstein–de Sitter universe became the favourite cosmological model. For a long time, the discrepancy between the observed and the theoretically required ratio of the matter density, responsible for the balance between the gravitational pull and the expansion velocity, was generously blamed on observational uncertainties. Towards the end of the twentieth century, observations of the microwave background and the time-dependent expansion rate of the Universe, determined from supernovae, made it clear that the Einstein–de Sitter cosmology had to be abandoned and the cosmological constant, Λ – frowned upon by Einstein, yet beloved by Lemaître and Eddington – was allowed back into cosmology.

Eddington's 'after dinner speech'

An amusing story about the Einstein–de Sitter paper is buried in a presentation by Eddington as part of a Cambridge 'Background to Modern Science' lecture series in 1936. Eddington's talk had the title 'Forty Years of Astronomy'. When it came to the expanding universe, he mentioned the joint paper by Einstein and de Sitter, and its attack on the cosmological constant. The story that Eddington relates would certainly have made a good after-dinner speech: 'It was a piece of mathematics, innocuous in itself, but put in such a way as to give the impression that these distinguished authorities had become sceptical about the cosmological constant. Einstein came to stay with me shortly afterwards, and I took him to task about it. He replied: "I did not think the paper very important myself, but de Sitter was keen on it." Just after Einstein had gone, de Sitter wrote to me announcing a visit. He added: "You will have seen the paper by Einstein and myself. I do not myself consider the result of much importance, but Einstein seemed to think that it was." ' (Eddington 1940, p. 128.)

15

Are the Sun and Earth older than the Universe?

At an early stage it was realised that the model of an expanding universe was laden with a heavy mortgage. The rate of expansion provides a first clue about the time that has elapsed since the beginning of expansion, which naturally would be identified with the beginning of the Universe. However, the expansion rate led to ages that were lower than the age of the Earth, and much lower than the age of the Sun, which at that time was estimated to be approximately 10^{12} years. Eliminating the discrepancies was a long process, which demanded a better understanding of stellar energy production, stellar evolution and a drastic revision of the Hubble constant.

The age of the Universe deduced from the expansion rate

The velocity–distance relation, found theoretically by Lemaître in 1927, and observationally by Hubble in 1929, says that there is a relation $v = H \times d$ for the velocity, v, between two galaxies and their distance, d. If no forces are acting, the expansion always runs at the same speed v, and the time, t, necessary to cross the distance d is $t = d/v = 1/H$. Thus, the inverse of the Hubble constant gives a first guess for the age of the expanding universe. In the 1930s, observations gave $H = 500\,(\text{km/s})/\text{Mpc}$, or $t = 1/H = 2 \times 10^9$ years.

When Einstein accepted the expanding universe in 1931, he categorically rejected the cosmological constant and, as we have seen in the last chapter, in collaboration with de Sitter he opted for a spatially flat, expanding universe with $\Lambda = 0$. In this model gravitation acts to slow expansion, and the age, τ, since the beginning of expansion is given by $\tau = (2/3) \cdot (1/H)$. This reduced the calculated age of the Universe to less than 1.4×10^9 years. Why should that bother the cosmologists?

The age of the Earth

In those years, the geological age of the Earth, as found from radioactive dating, was estimated to be at least 1.5×10^9 years, but ages up to 3×10^9 years and even higher were advocated (Brush 1996). That did not bode well for a universe of less than 2×10^9 years. Worse was to come from stellar astrophysics.

The age of the Sun and the stars

It was generally agreed that the Sun had to be at least as old as the Earth. Fossils suggested that the Sun must have radiated with approximately the same brightness for hundreds of millions of years. The source of stellar energy became a much-debated issue. Infall of comets on to the Sun or gravitational contraction had already been eliminated in the nineteenth century. It was shown that comet infall had been based on exaggerated estimates for the mass of the comets. With the successive lengthening of geological timescales, gravitational contraction of the Sun was falling short of the original expectation. Contraction to the present size could have provided fuel for at most 20 million years whereas, from sedimentation rates, nineteenth-century geologists estimated the age of the Earth to be several hundred million years.

Help came from Einstein. In 1905 he had published his most famous formula: $E = m \times c^2$. This led to speculations that stellar energy could be produced by an annihilation process, whereby mass was converted to energy.

Jean Perrin, 1926 Nobel Prize winner, suggested in 1919 that stellar luminosities could be sustained by the fusion of lighter atoms, such as helium, into heavy atoms, such as radium. In this way, Perrin said, 'it is the disappearance of potential electric energy, and not gravitational energy, which pays for the prodigous expenditure of stellar radiation.' (Perrin 1919, p. 91, translated from French.)

Similar ideas were developing across the Channel. In 1919, James Jeans, who together with Eddington was then the most renowned British astrophysicist, wrote a book on *Problems of Cosmogony and Stellar Dynamics*. Timescales were one of his worries. According to his sources, the age of the Earth is about 250 million years. But gravitational energy gained by the Sun in contracting to a homogeneous mass of its present size would only represent radiation for 18.3 million years. Jeans cannot name any other viable source of energy. But: 'There remains, apparently as a last resource, the possibility of energy being created by the destruction of matter, as for instance by positive and negative charges rushing together and annihilating one another.' The reduction of the Sun's mass by only one per cent would furnish radiation at the present rate for 150 000 million years (Jeans 1919, p. 286).

In the early decades of the twentieth century, stellar evolution became an astronomical topic. Hertzsprung and Russell had independently found that in a plot of absolute magnitude against surface temperature or spectral class, most stars fell roughly along a line called the *main sequence*, while a few fell in another region labelled *giant sequence*. Eddington had theoretically shown that there is a tight relation between the mass of a star and its luminosity, the most massive being the most luminous.

By 1920, it became generally accepted that a star formed when a proto-stellar nebula condensed towards a stable gaseous sphere. It was believed that when condensation reached the stable point, it had also reached the hottest stage in the life of the star. It was thought that during its evolution, the star was running down the main sequence from hot to cool, losing mass due to radiation (e.g. Eddington 1920). The diminution of mass, Δm, would correspond to the energy, ΔE, radiated into interstellar space, according to $\Delta m = \Delta E / c^2$. Comparing mass differences and luminosities, the amount of energy radiated per unit time, $\Delta E / \Delta t$, along the Hertzsprung–Russell sequence would then provide the time, Δt, that had elapsed between two stages in the life of the star.

Condon calculated the lengths of such tracks. If we want to know the age of the Sun, we would have to know its initial mass. However, for an order of magnitude estimate the uncertainty of the initial mass is not crucial. If it had started at ten times its present mass, its age would be 9×10^{12} years, but only 3×10^{12} years if it had begun with 1.5 times its present mass. However, a star like our Sun would still have a future life of 25×10^{12} years (Condon 1925).

In 1928, Jeans returned to the source of stellar energy. Annihilation was still the watchword. 'The principle of this calculation is very simple. When a proton and an electron of masses m, M neutralise one another in the sun and disappear, the sun loses material mass of amount $m + M$ ... Hence, in accordance with Einstein's principle already explained, the energy of the radiation must be $(m + M)c^2$.' (Jeans 1928, p. 109.) Because it was supposed that the Sun began its life with a much higher mass than it has now, its total lifespan is millions of millions of years.

This would have been fine for Einstein's static universe. However, when the fashion shifted to the expanding universe after 1930, everyone agreed that timescales were a problem. The few times 10^9 years that had elapsed since the beginning of expansion were not the stuff to impress the sceptics, and Dingle lamented that 'the Universe of nebulae appears to be many times younger than the stars which compose it' (Dingle 1933a).

Progress in atomic physics, nuclear physics and the physics of stellar interiors would lead to a different concept of stellar evolution over the following 30 years. A first advance concerned elemental abundances. Whereas, up to 1928, the astrophysical big shots were of the opinion that the chemical composition of the Sun and other stars was similar to that of the Earth, the spectral analysis of Cecilia Payne-Gaposchkin and Albrecht Unsöld showed that the majority of atoms in the stars are hydrogen, and the heavier elements are in the minority. This gave rise to the suspicion that originally all matter had been in the form of hydrogen, and that some of it was later converted into helium and heavier elements. Thus was born the idea that the elements heavier than hydrogen were created inside the stars. There was still the opinion that about 40% of the atoms in the Sun were heavier than hydrogen and had thus been produced by fusion. It was only in the 1950s, when progress in astrophysics led gradually to revised proportions of approximately 92% hydrogen (H), 8% helium (He) and 0.1% heavier elements (in number), that stellar abundances stopped being a threat to the expanding universe. For a commented compilation of earlier abundance determinations, see Sandage (2000).

Strömgren introduced new ideas about stellar evolution around 1933. He suggested that stars might initially have consisted mainly of hydrogen. Transformation of hydrogen into more complex elements would then provide the energy, which was radiated away. That process would not change the mass by much, but it would change the chemical composition and make the star expand (Strömgren 1933). Red giants were therefore not a stage in the contracting proto-stellar nebula, but a late stage in the life of a star. The very long timescales of stellar evolution shortened when it was realised that no complete annihilation took place: during fusion, less than one per cent of the mass is turned into energy. This also reduced the discrepancy between the age derived from stellar evolution and that derived from expansion. It might be added that nuclear physics proper only began after the discovery of the neutron in 1932.

Baade and Sandage drastically reduce the Hubble constant

In the mid 1960s, the value of the Hubble constant dropped dramatically to about one seventh of Hubble's estimate. This was due to Baade's recalibration of the Cepheid distance scale (Baade 1956) and Sandage's discovery that the knots identified by Hubble as the brightest stars in more distant galaxies were really H II regions. In 1958, the combination of the two effects reduced the Hubble constant to $H \approx 75$ (km/s)/Mpc (Sandage 1958). This went a considerable way to eliminating the age problem. A still better understanding of the generation of the elements and of stellar evolution did the rest.

16

In search of alternative tracks

Does cosmology really need General Relativity? And if so, is the expanding universe the only possible explanation for the observed redshifts? When, in 1931, Hubble and Humason confirmed the approximately linear redshift–distance correlation by their extended observations, relativistic cosmology definitely ceased to be a purely theoretical exercise. However, it was not surprising that attempts were made to explain the observed redshifts within more conventional physics. After all, was it really necessary to abandon the age-old concept of Euclidean space, as demanded by Einstein's relativistic model of 1917, with its cosmological term, which had no relation to any known form of energy? And de Sitter's universe was even more abstruse: it is empty, and time is running differently at different locations. But it had to be admitted that General Relativity explained the perihelion shift of Mercury and the bending of light in eclipse observations. Yet, Einstein's cosmological model produced no redshifts, and de Sitter's was incomprehensible.

We look at two attempts to explain redshifts without the expanding universe of General Relativity. They were by Zwicky and Milne. In addition, we shall briefly mention the Steady State alternative.

Zwicky and the gravitational drag

When Hubble and Humason published their observations in 1929, they refrained from interpretation; this, they said, was for the theoreticians. Zwicky responded. He was working at the California Institute of Technology in Pasadena and knew the astronomers of Mt Wilson. He also knew and accepted General Relativity; from the lack of citations we must assume that he was unaware of Friedmann and Lemaître in 1929.

Zwicky saw in the gravitational 'Drag of Light', as he called it, a chance to account for the redshifts. He explained: 'According to the relativity theory, a light quantum hν has an inertial and a gravitational mass $h\nu/c^2$. It should be expected, therefore, that a quantum hν passing a mass M will not only be deflected but it will also transfer momentum and energy to the mass M and make it recoil.' During this process, the light quantum will change its energy and, therefore, its frequency. For the mean density of matter, ρ, Zwicky referred to Hubble's information that it was between 10^{-31} and 10^{-26} (g/cm^3). For these densities, he calculated a frequency shift $\Delta\nu/\nu$ between 3×10^{-7} and 3×10^{-2} due to the gravitational drag. Hubble's observations gave $\Delta\nu/\nu \approx 1/600$. This convinced Zwicky that 'a further elaboration of the theory seems to be worthwhile' (Zwicky 1929).

In 1933, Zwicky returned to the subject (Zwicky 1933). He had been studying clusters of galaxies. He worried about the large deviations of redshifts of individual galaxies from the mean redshift of the whole cluster. In addition, he still considered that van Maanen's rotational velocities in individual galaxies had to be explained as well. This, he thought, created a serious problem for the general interpretation of the observed redshifts. In the same publication, Zwicky pointed to the likely existence of dark matter (Zwicky 1933, p. 125).

Zwicky accepted that there were two viable suggestions for the explanation of the observed redshift: (1) the general expansion of the Universe, as suggested in the Einstein–de Sitter model, and (2) his own suggestion of an interaction between the radiation from distant galaxies and intergalactic matter. But he concluded that none of the proposed explanations satisfy. In his opinion both rested on very hypothetical foundations.

Zwicky remained sceptical about the expanding universe. He argued that if one started from an originally random distribution of nebulae, its timescales – the cluster crossing times – were too short to allow the formation of stationary clusters of galaxies (Zwicky 1935). However, his arguments did not find many followers.

Milne's static Euclidean space

With the acceptance of Lemaître's expanding universe by Eddington and de Sitter in 1931, cosmologists thought they had a theory that, at least in principle, could explain the present, and even predict the future of the Universe. Hardly had a consensus about the new cosmology been reached, when it was challenged by a quite different mechanism of expansion and cosmic evolution. E. A. Milne, Professor of Mathematics at the University of Oxford, had taken the initial step and remained one of the main proponents. He thought

At greater distances from the centre than r_0 a sorting process goes on, the particles dividing themselves into spherical zones of gradually increasing recession-velocity as we go outwards. The mean velocity of recession at

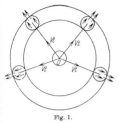

distance r at time t is approximately r/t. For for t sufficiently large, the distance from the origin of a particle of velocity V is approximately V. To see this in more detail, consider the sub-group of particles which are moving with given vector-velocity V. At time $t = 0$ they occupy a sphere centre O, of radius r_0. At time $t > 0$ the sphere of particles will have moved a vector-distance $R = Vt$. For $t > r_0/|V|$ this sphere will be

Fig. 1.

between the spheres centre O, of radii $|V|t - r_0$ and $|V|t + r_0$. Thus

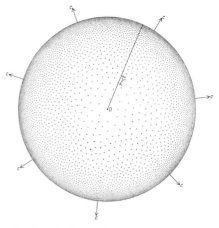

Fig. 16.1 Milne's velocity distribution for galaxies. An alternative explanation for the observed nebula redshifts. In Milne's model the density is increasing outwards, at first slowly, ultimately to infinity. (Left: Figure 1 from Milne 1933. Right: Plate 1 from Milne 1935.)

that the old Newtonian concepts, supplemented with Special Relativity, would suffice. 'The phenomenon of the expansion of the universe is shown to have nothing to do with gravitation, and to be explicable, qualitatively and quantitatively, in terms of flat, infinite static Euclidean space' is the first sentence of the provocative article of 95 pages, submitted in November 1932 from the Potsdam Albert-Einstein-Institut (Milne 1933). Milne had spent the summer there, but said in the paper that most of the work was carried out at Oxford in May 1932.

The Milne universe is created at time zero out of a very small space. The nebulae were originally concentrated in a very small volume. They have since been radially expanding according to the individual velocities of each nebula. They expand into a flat (Euclidean) spacetime. All nebulae with a given velocity v are now at position $r = v \times t$, where t is the time since the beginning. The largest velocity approaches the speed of light. Therefore the galaxies occupy the interior of a bubble that expands at the speed of light into previously empty space. Milne explained that if an assembly of objects were released from a point-like location with almost any arbitrary velocity distribution, after a time, to an observer on any one of these objects, the behaviour of the rest of the objects would look like the expanding universe. They can be treated as test particles; gravitation is ignored. The closer they come to the speed of light, the nearer they will be to the surface of the bubble.

Milne accepted Special Relativity. When galaxies shoot out at different speeds, but close to the speed of light, the Lorentz contraction becomes noticeable. Nebulae receding with high velocities drop appreciably in brightness to an observer at rest, and are in fact much nearer than would be estimated from their apparent brightness. There are many questions to answer: why is every galaxy endowed with a certain velocity; what is the significance of an initial configuration; what about the preferential region of the initial configuration; what is the significance of the time, $t=0$? Milne provides the answers in his 95 pages.

The reaction followed immediately. Herbert Dingle responded in May 1933 with a critical analysis of 13 pages (Dingle 1933b). He agreed that observationally it was not yet possible to distinguish between Milne's and the 'expanding space' theory. He found Milne's assumptions in general less convincing and reproached him that its 'fundamental view-point is inconsistent with the general principles of scientific thought'. Eddington pointed out a particular problem in his public lecture to the IAU in September 1932. 'To provide a moderately even distribution of nebulae up to 150 million light years' distance, high speeds must be very much more frequent than low speeds; this peculiar anti-Maxwellian distribution of speeds becomes especially surprising when it is supposed to have occurred originally in a compact aggregation of galaxies.' (Eddington 1933, p. 64.)

Milne's model violates basic principles of General Relativity. In particular, it rejects the principle that mass and energy curve spacetime. It does not accept expansion of space, but explains the observed redshifts by recessional velocities, associated with the initial explosion. As theory and observations progressed, it became obvious that Milne's model could not account for essential observational facts: it does not explain the cosmic microwave background, although it is a kind of Big Bang scenario. Whereas the abundances of light elements are a strong piece of evidence for the Big Bang, Milne had nothing to offer in that respect. The mean density of galaxies also speaks against Milne. Observations show a mean density of galaxies that is constant with distance. To produce such a distribution with the Milne's model would need a very peculiar initial velocity distribution.

The main protagonists in the debate were Milne, Eddington, Dingle (at that time secretary of the Royal Astronomical Society), McCrea, McVittie, Robertson and Walker. Allegiances with the two views changed in the course of time. The discussions turned into a general philosophical debate that spilled over into the debate on the Steady State universe. It died down in the 1950s. Dingle, in his presidential address to the Royal Astronomical Society, probably thought he had found an appropriate epitaph: 'Even idle speculation may not be quite valueless if it is recognized for what it is. If the new cosmologists would observe this proviso, calling a spade a spade and not a perfect agricultural principle, ones only cause for regret would be that such great talents were spent for so little profit.' (Dingle 1953, p. 406.)

Although Milne's idea had its supporters, it never seriously jeopardised the concept of the expanding universe.

The Steady State alternative

Jeans' preliminary Steady State speculation

Some preliminary thoughts about a Steady State universe, with matter being continuously created, can already be found in Jeans' textbook, *Astronomy and Cosmogony*, published in 1928. At the end of Chapter XIII on 'The great nebulae', he muses about the failure to find a satisfactory explanation for the spiral arms. He then says: 'Each failure to explain the spiral arms makes it more and more difficult to resist a suspicion that the spiral nebulae are the seat of types of forces entirely unknown to us, forces which may possibly express novel and unsuspected metric properties of space. The type of conjecture which presents itself, somewhat insistently, is that the centres of the nebulae are of the nature of "singular points," at which matter is poured into our universe from some other, and entirely extraneous, spatial dimension, so that, to a denizen of our universe, they appear as points at which matter is being continually created.' (Jeans 1928, p. 352.)

Jeans returned to continual creation at the end of his book: 'It is difficult, but not impossible, to believe that matter can be continuously in process of creation, or possibly of re-creation out of stray radiation.' (Jeans 1928, p. 413.) He argued that we still are free to think that astronomical bodies are passing in an endless steady stream from creation to extinction, just as human beings, with new generations ready to step into the place vacated by the old. But the number of objects in the various stages of development did not seem to exist in proportion to the times taken to pass through these stages – even roughly. Thus, the cosmogonist can only end with a question 'others, more confident or more fortunate, may, if they wish, attempt an answer'.

Bondi, Gold and Hoyle

The Steady State theory was the most serious challenge to the evolving Friedmann–Lemaître type of universe. It was originally presented in Edinburgh in 1948, at the first meeting of the Royal Astronomical Society ever to be held away from London. Because the new theory was of such striking novelty – the continuous creation of matter – the three authors became famous overnight. Its initial content is given in two publications that appeared in the *Monthly Notices* of 1948 (Bondi and Gold 1948, Hoyle 1948). They were the result of a close collaboration between the three authors, all at Cambridge.

Bondi, Gold and Hoyle had aesthetic as well as very practical objections to the Einstein–de Sitter and Eddington–Lemaître models. The practical objection was the age paradox, derived from the high value of the Hubble constant, which in 1948 was still accepted as ~500 (km/s)/Mpc. The aesthetic offence was the restriction of homogeneity to space; time should be included as well. In addition, they wanted to be rid of the cosmological constant.

Their basic assumption was the 'perfect cosmological principle' – Hoyle called it the 'wide cosmological principle'. It assumes that the Universe is not only spatially homogeneous, but also temporally unchanging on the large scale. The version of Bondi and Gold differed from Hoyle's in detail, although they agreed on the general concept. Hoyle outlines how the Steady State theory should overcome the problems mentioned above: 'Using continuous creation of matter, we shall attempt to obtain, within the framework of the general theory of relativity, but without introducing a cosmical constant, a universe satisfying the wide cosmological principle that shows the required expansion properties and in which localized condensations are continually being formed.' (Hoyle 1948.) The Steady State theory is built on the hypothesis of continuous creation. Although new matter enters the Universe, local properties remain constant and, in spite of permanent creation, there is no qualitative evolution.

The essential step in Hoyle's formalism was the introduction of the tensor C into the Einstein field equations. This tensor played a role similar to that of the cosmological constant. However, the time component C_{00} was set to zero. This allowed Hoyle to construct a de Sitter model with the big difference that it contained matter. His formalism gave him a redshift and a continuous creation of matter.

To an observer, galaxies in the Steady State universe show the same distribution and spectral shifts as in the de Sitter expanding solution. There exists a radius of the observable universe. The total amount of matter within that range remains constant. The newly created matter compensates for the loss of matter out of this volume due to expansion. Hoyle remarks that with their knowledge of nuclear physics it was not possible to make a statement on the identity of the created particles. Neutron creation appears to him to be the most likely possibility. Subsequent disintegration might supply the hydrogen required by astrophysics; moreover, the electrical neutrality of the Universe would be guaranteed.

Thus, the Steady State universe is homogeneous in space as well as in time. It is stationary but not static, it has no beginning and no end. As matter is continuously created throughout the Universe, it can be perpetually expanding. We have an eternity behind and an eternity in front of us. As physical conditions are always the same, there is a certain probability of repetition of events.

The Steady State theory was much debated in the 1950s, particularly in Great Britain. It did much to keep the cosmological debate alive. As an unquestioned advantage it had no age paradox. But when, in 1958, the work of Baade and Sandage lowered the Hubble constant by a factor of approximately seven, the age problem of the classical expanding universe began to disappear.

According to Sandage, the Pasadena astronomers had a simple but very pertinent argument against the Steady State theory: 'The Pasadena astronomers never believed in the Steady State model because of their conviction that all galaxies are the same age, whereas in Steady State new galaxies must be forming continuously.' (Sandage, personal communication, November 2007.)

The decisive blow against the Steady State theory was struck in 1965 with the discovery of the microwave background radiation. This could be easily explained by the hot Big Bang, but not by Bondi, Gold and Hoyle.

17

The seed for the Big Bang

Why does the Universe expand? Eddington and Lemaître were the first to delve into that subject. Although Lemaître had neglected radiation pressure when deriving the redshift–distance relation in 1927, he touched on it at the end of his article: 'It remains to find the cause of the expansion of the universe. We have seen that pressure of radiation does work during the expansion. This seems to suggest that the expansion has been set up by radiation itself.' (Lemaître 1927, 1931a.)

When Eddington showed that Einstein's static universe was unstable, he suggested that such a pseudo-static universe might have been the original status of the Universe, out of which the expanding universe was born (Eddington 1930). But Eddington and Lemaître differed fundamentally in their view about the beginning. Eddington and his students and collaborators persisted in an expansion out of an unstable static state, whereas Lemaître favoured an explosive beginning. This could be either a true beginning out of a primeval atom, or the turning point in a cyclic history of the Universe.

When it came to discussing the beginning, Lemaître was in a delicate position. As a priest he was under suspicion of trying to smuggle theology into astrophysics. However, he had no such intentions. On several occasions he made it clear that he did not see the Bible as a textbook of astronomy, and expected scientific truth to emerge out of scientific discussions.

Because we are restricting ourselves to the discovery phase of the expanding universe, we shall not follow up the whole debate about the origin of the Universe – which is still going on – but just mention the first few ideas published on this subject. The interlude, where the Big Bang was in competition with the Steady State universe, was discussed at the end of the previous chapter.

Expansion out of Einstein's static state

The *Monthly Notices* of 1930 and 1931 saw a lively exchange on how the pseudo-static Einstein universe would slide into expansion. How would disturbances affect the unstable equlibrium of Einstein's model, and what would trigger expansion? Eddington left no doubt about the following point: 'The initial small disturbance can happen without supernatural interference' (Eddington 1930). Was that an allusion to Lemaître? Then he suggested that an originally uniformly diffused nebula condensed into galaxies via gravitational instabilities, and that this evolutionary process started off the expansion. Alternatively, the initial equilibrium might have been upset by the conversion of matter into radiation; Tolman's paper about conversion of matter into radiation had just appeared. However, Eddington found that this process would cause contraction, and had therefore to be rejected.

McCrea and McVittie joined the discussion. The two newcomers first thought they had demonstrated that 'if the actual universe started as an Einstein universe and is now expanding, as the work of Lemaître and Eddington suggests, this expansion cannot have been caused by a redistribution of matter into massive nuclei. Thus, as yet, no mechanism for setting up the expansion has been found.' (McCrea and McVittie 1930.) But, when McVittie had another look at the subject, he found that although a single condensation would be followed by a contraction, a finite number of small condensations would result in an expansion (McVittie 1931).

Also Lemaître looked at the condensations advocated by Eddington, McCrea and McVittie. No, is his answer, these condensations do not affect the radius of the Universe at equilibrium. He suggested a different mechanism for starting the expansion. After lengthy calculations he found: 'If, in a universe in equilibrium, the proper mass begins to vary, the radius of the Universe varies in the same sense. If, in a universe in equilibrium, the pressure begins to vary, the radius of the Universe varies in the opposite sense. Therefore stagnation processes induce expansion.' By stagnation, Lemaître meant the following: when there is no condensation, a fair fraction of the energy may be able to wander freely through the Universe, e.g. in the form of radiation. But when condensations are formed this free energy can be captured by the condensations and then remain bound to them. Stagnation is a diminution of the exchange of energy between distant parts (Lemaître 1931b).

A few months later, McVittie and McCrea retracted their statements because they had not paid sufficient attention to conditions at infinity. At the same time, they criticised Lemaître's approach. Thus the discussion about triggering the expansion was still wide open. However, they judged the whole debate to be rather academic. After all, Einstein's universe was not in a stable equilibrium, and it was therefore unnecessary to twist one's brain to find a way to get it out of equilibrium – it will happen anyway (McCrea and McVittie 1931).

Expansion out of a primeval atom: The ancestor of the Big Bang

On 5 January 1931, Eddington delivered the presidential address to the Mathematical Association. Its title was 'The End of the World'. He expressed the opinion that the Universe had always existed: 'Philosophically, the notion of a beginning of the present order of Nature is repugnant to me' (Eddington 1931).

Lemaître responded immediately. To Eddington's 'The End of the World', he countered with 'The Beginning of the World'; well, the full title was 'The Beginning of the World from the Point of View of Quantum Theory' (Lemaître 1931c). The title leaves no doubt; Abbé Lemaître was not going to preach theology, he intended to treat the beginning of the Universe as serious science. He saw nothing repugnant in the concept of a beginning. His short letter to *Nature* contains the first suggestion for what was later termed the Big Bang.

Lemaître replaced the singularity of time zero by a single atom that contained all the matter and energy. He saw an analogy to radioactivity: one complex atom decays into several less complex atoms. Why should there not have been a single, exceedingly complex atom, containing all the matter of the Universe? 'This highly unstable atom would divide in smaller and smaller atoms by a kind of super-radioactive process.' Along this line, he argued, if we go back in time we must find fewer and fewer quanta, until we find all the energy of the Universe packed into a few or even in a unique quantum. And, he continues: 'If the world has begun with a single quantum, the notion of space and time would altogether fail to have any meaning at the beginning; they would only begin to have a sensible meaning when the original quantum had been divided into a sufficient number of quanta. If this suggestion is correct, the beginning of the world happened a little before the beginning of space and time.' (Lemaître 1931c.)

Radioactive decay as a cosmic phenomenon was in the air, as can be seen in a lecture given to the British Association in the autumn of 1931 by Sir James Jeans (Jeans 1931). However, Lemaître placed it within the concept of the expanding universe. His single, complex atom was a highly ordered 'universe'. The beginning is therefore also the moment when the entropy of the Universe is zero, and all the matter is condensed in a near-singularity. The decay of this initial atom would see a growth in entropy, with the Universe evolving towards a state of infinite entropy.

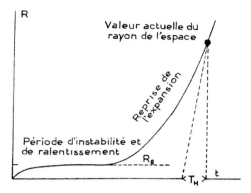

Fig. 17.1 The beginning according to Lemaître. A first expansion is due to the decay of the primeval atom. It leads to a stagnation period of a quasi-static Einstein universe at R_E. However, Λ being slightly above the equilibrium value, expansion is resumed. T_H is the age of the Universe when calculated from the presently observed rate of expansion; it essentially corresponds to the Hubble time, which is the inverse of the Hubble constant. Due to the long period of stagnation, this is much shorter than the real age of the Universe. We show an original presentation by Lemaître. (Lemaître 1947.)

Later in the chapter, we shall see that Eddington remained unhappy with the concept of an abrupt beginning. Thus, there were two fundamentally different ideas about the origin of our Universe. Lemaître saw it as a transmutation from a primeval highly condensed state, which in itself contained the seeds of subsequent evolution. The breaking up of that state leads to the creation of the elements, of space and time, and unleashes the expansion of space. In contrast, Eddington wants no beginning and no creation. The Universe has been in a state of equilibrium, out of which it was woken into expansion by gravitational instabilities. Both proponents argue on a purely scientific basis, but from different philosophical inclinations. Lemaître's concept is closer to our modern view than that of Eddington. Lemaître counted on the progress in quantum theory to shed light on the primeval atom; today we speculate on phase transitions and look for help from high-energy physics.

Lemaître is also troubled by a philosophical question, to which he offers a scientific answer: is fate written into this primeval atom, and is the whole course of evolution already determined at the very beginning? No, he says, the principle of indeterminacy prevents that. The disintegration of the primeval atom can therefore happen in many different ways. The universe that emerges may well be a very improbable one among all the possibilities. 'The whole story of the world need not have been written down in the first quantum like a song on the disc of a phonograph. The whole matter of the world must have been present at the beginning, but the story it has to tell may be written step by step.' (Lemaître 1931c.) – Thus free will is saved.

A case for the cosmological constant

For Lemaître, the cosmic evolution after the decay of the primeval atom is due to an imbalance between two opposing cosmic forces: gravitation and dark energy. Dark energy is the modern term. In the 1930s, it was embodied in the cosmological constant, Λ; at the end of this chapter we shall see that Lemaître equated it with vacuum energy. The cosmological constant played a crucial part in all these discussions. When Einstein dissociated himself from Λ in April 1931, neither Lemaître nor Eddington followed suit. On the contrary, both considered Λ was an essential element of the theory.

During the 1932 meeting of the International Astronomical Union at Cambridge (Massachusetts), Eddington delivered a public lecture on 'The expanding Universe'. An extended version of that talk was published as a book (Eddington 1933). In its preface, Eddington calls Λ the 'hidden hand' in the story of expansion, and on page 24 he says: 'Not only does it unify the gravitational and electromagnetic fields, but it renders the theory of gravitation and its relation to space-time measurement so much more illuminating, and indeed self-evident, that return to the earlier view is unthinkable. I would as soon think of reverting to Newtonian theory as of dropping the cosmical constant.' And he reiterates his creed near the end of the book (page 104): 'To drop the cosmical constant would knock the bottom out of space.'

There also existed a very practical reason for retaining Λ. It allowed augmentation of the age of the Universe. The expansion rate of approximately $H = 500$ (km/s)/Mpc was generally accepted. For the spatially flat Einstein–de Sitter model, with a cosmological constant $\Lambda = 0$, the result was an age of approximately 1.3×10^9 years. To avoid the paradoxical case of the Earth being older than the Universe, both Eddington and Lemaître put their hope on a slow way out of an Einstein pseudo-static state with a radius, R_E, and a corresponding cosmological constant, Λ_E.

The evolution of the radius of the expanding universe, R, away from the equilibrium value, R_E, and its dependence on time, t, is connected by a relation that depends logarithmically on the difference between R and R_E: $t \sim \log((R - R_E)/R_E)$. Going into the past, $t \to -\infty$, and the difference $R - R_E$ approaches zero. Thus, if Nature had chosen a size R just slightly above R_E, it could have remained in that state for a very long time, evolving only very slowly to a larger R. However, Eddington was not happy with a logarithmic infinity. He considered it unlikely that Nature would creep out of equilibrium that slowly once the process had started. Tolman elaborated further on the theoretical and computational details of these processes (Tolman 1934, § 157).

For Lemaître, the pseudo-equilibrium state became a transitional stage. With a positive curvature, and Λ slightly above the equilibrium value Λ_E, the history of the Universe crosses three qualitatively different periods: (1) a rapid initial expansion due to the decay of the primeval atom, (2) a stagnation period of very slow expansion, resembling Einstein's equilibrium universe, and (3) our present expansion, where the repulsive force of the cosmological term dominates. When building models, the first stage would just need to expand slightly above the equilibrium state to make the Universe slide into a pseudo-Einstein state. This could then provide sufficient time for the formation of stars. (Also see Lemaître 1967.)

Lemaître realised that the decay of the primeval atom would have liberated enormous amounts of energy, and that there should be some relics still floating through space. He suspected cosmic rays to be such a relic, and associated them with the formation of the first stars from the decaying primeval atom during the first episode of expansion (Lemaître 1931d, p. 407, 1931e). However, at the time very little was known about cosmic rays, and even less about star formation. Thus, he was quite entitled to be somewhat vague about the details. About the origin of the primeval atom, he remained silent.

Is there a beginning to the Universe?

Lemaître was much interested in the initial period. He envisaged two alternatives: a cyclic universe and a unique edition. Friedmann had already mentioned the possibility of a cyclic universe. Lemaître depicts how the present expansion might have been preceded by a contraction, and the annihilation of the former structure during the phase of infinite curvature. He calls this 'l'univers phénix', where the present universe would have started with matter inherited from the past universe: a Big Crunch followed by the Big Bang. But he also envisaged a universe where the departure from time zero was indeed a real beginning out of an initial atom.

Eddington did not agree. In his 1932 address to the IAU General Assembly, he gives us a condensed description of his vision:

> Views as to the beginning of things lie almost beyond scientific argument. We cannot give scientific reasons why the world should have been created one way rather than another. But I suppose that we all have an aesthetic feeling in the matter.

Then, he shares his aesthetic feelings with us:

> Since I cannot avoid introducing this question of a beginning, it has seemed to me that the most satisfactory theory would be one which

made the beginning not too unaesthetically abrupt. This condition can only be satisfied by an Einstein universe with all the major forces balanced. Accordingly the primordial state of things which I picture is an even distribution of protons and electrons, extremely diffuse and filling all (spherical) space, remaining nearly balanced for an exceedingly long time until its inherent instability prevails. We shall see later that the density of this distribution can be calculated; it was about one proton and electron per litre. There is no hurry for anything to begin to happen. But at last small irregular tendencies accumulate, and evolution gets under way. The first stage is the formation of condensations ultimately to become the galaxies; this, as we have seen, started off an expansion, which then automatically increased in speed until it is now manifested to us in the recession of the spiral nebulae.

Then he leaves us to make our own choice: 'I cannot but think my "placid theory" is more likely to satisfy the general sentiment of the reader; but if he inclines otherwise, I would say – "Have it your own way. And now let us get away from the Creation back to problems that we may possibly know something about." ' (Eddington 1933, pages 55–60.)

Is the Big Bang the origin, or a transition in a cyclic universe?

In an article of 1933, published in French, Lemaître attacks the problem 'de l'évanouissement de l'espace' – the disappearance of space in a cosmic singularity. In the contraction phase of Friedmann's cyclic universe, the radius could shrink to zero. But what about the Schwarzschild radius? Lemaître shows mathematically that the singularity of the Schwarzschild radius is not a real singularity, but can be eliminated by a different choice of coordinate system (Lemaître 1933, p. 80). What then is the smallest radius of the Universe? When all the matter and stars of the Universe are concentrated in a single mass, it will attain an enormously high temperature. But even a degenerated gas will not be the ultimate state. 'When distances between the atomic nuclei and the electrons are of the order of 10^{-12} cm, the non-Maxwellian forces which prevent the mutual penetration of the ultimate particles will become predominant and will doubtless be capable to stop further contraction. The universe would then be comparable to a colossal atomic nucleus. When contraction has been arrested, the process has to restart in the opposite direction.' (Lemaître 1933, p. 85.)

Thus, we are back at the primeval atom. Lemaître then adopts Eddington's speculative value for the total number of protons in the Universe: 10^{78}. This gives him a radius of 10^{14} cm for the smallest size of the Universe. Thus it would have easily fitted within the Solar System and inside the orbit of Saturn. From a cosmological point of view, this resurrection out of a near-zero space is a beginning, in the sense that every previous structure has been completely wiped out. However, from the value of the Hubble constant, Lemaître concludes that the length of such a cycle is only about two billion years, which is less than the age of the Earth. The cyclic solution has therefore to be excluded. This, for aesthetic reasons, he very much regrets: 'Ces solutions où l'univers se dilatait et se contractait successivement en se réduisant périodiquement à une masse atomique des dimensions du système solaire, avaient un charme poétique incontestable et faisaient penser au phénix de la légende.' (These solutions, where the Universe expanded and contracted in succession, reducing itself periodically to an atomic mass of the size of the Solar System, had an incontestable poetic charm, and recalled the legendary phoenix.)

What a lovely example of what Eddington called *an aesthetic feeling*. If given the choice of whether our Universe was the first and only one that ever was, or whether it was one in a succession of universes, Lemaître would be poetically charmed to be part of a link in a cyclic cosmos. However, with a slight regret, he accepted living in a first edition.

Dark energy: Lemaître equates the cosmological constant Λ with vacuum energy

Lemaître continued to think about possible causes for the expansion, and he believed he found it in what today we call dark energy. In a paper, read on 20 November 1933 before the National Academy of Sciences, Lemaître associated the cosmological constant Λ (Lemaître calls it λ) with vacuum energy: 'Everything happens as though the energy in vacuo would be different from zero. In order that absolute motion, i.e., motion relative to the vacuum, may not be detected, we must associate a pressure $p = -\rho \cdot c^2$ to the density of energy $\rho \cdot c^2$ of vacuum. This is essentially the meaning of the cosmological constant λ which corresponds to a negative density of vacuum ρ_0 according to $\rho_0 = \lambda c^2 / (4\pi G)$.' (Lemaître 1934.) To our knowledge this is the first time that, via the cosmological constant, the expanding universe was related to vacuum energy.

Schrödinger had considered a universe in equilibrium without a cosmological term in 1918 (Schrödinger 1918). He showed that if the energy–momentum tensor is modified, Einstein's cosmological model can be obtained without introducing the cosmological term Λ. Pressure has to be replaced by an expression that contains energy and curvature. In modern terminology, this corresponds to vacuum energy; Schrödinger did not, however, explicitly mention this. Einstein responded to his article (CPAE 2002, Doc. 3), and the same theme also appeared in an exchange of letters between Klein and Einstein, where singularity-free solutions of de Sitter's model were the main subject (CPAE, 1998, letters 566, 567). The essence was that the cosmological term could indeed be replaced by a constant pressure term. To this end, negative pressure is added to the energy–momentum tensor, such that $p \sim -\Lambda$ or $p \sim -1/(\kappa R^2)$, where $1/\kappa$ corresponds to the total energy per volume, and R is the radius of curvature. However, none of these authors gave an explicit physical meaning of cosmological consequence to that pressure.

Seitter and Duemmler (1989) and Straumann (2007) give further information on the history of vacuum energy and its relation to gravitation and cosmology.

18

Summary and Postscript

To find the way through the history of science one usually singles out groundbreaking discoveries and their discoverers; we shall do the same. But we should remember that every discovery happens in a specific context and with its own particular historical background. It is usually more complicated, and the product of a collective effort involving many more people than we mention in our text.

A brief recollection

Europe becomes acquainted with astronomy

After a few preliminaries on Greek astronomy, we began our journey with Sacrobosco who, in the middle of the thirteenth century, wrote *De Sphaera*, an extraordinarily successful astronomical textbook. It was instrumental in shaping the ideas on astronomy and cosmology right into the seventeenth century. The book is based on Ptolemy and his Islamic commentators, and teaches the geocentric, finite, spherical universe, which turns eternally around the immobile Earth. When, in 1543, Copernicus shifted the centre of the Universe to the Sun, he retained Ptolemy's spherical universe, bounded by a finite stellar shell. At that time, the only objects known in the sky were the planets, the Moon, the Sun and the fixed stars. Tycho Brahe added comets; previously, they were thought to be atmospheric phenomena.

Discovering the first nebulae

The first known European description of a galaxy – the great nebula in Andromeda – was published in 1614 by Simon Marius. In 1612, he had discovered it through the telescope. Andromeda was also included in the beautiful celestial catalogue of Hevelius, published in 1687. However, this nebula had already been mentioned in the tenth century by Abd ar-Rahman as-Sufi, also known as al Sufi, who lived in Isfahan at the court of the emir Adud ad-Daula.

In 1716, nebulae were the subjects of a report by Halley to the Royal Society; Andromeda was one of them. Halley's presentation bestowed respectability on this new class of astronomical objects and, by having them confirm the message of Genesis, he added a mystical touch to them.

No supernatural connotation was involved in the grand design of Kant and Laplace, who saw in nebulae large stellar agglomerates, similar to our Milky Way. With Newton's law of gravitation they believed they had found the major force responsible for cosmic evolution. Thus, at the end of the eighteenth century, the concept of island universes – today's galaxies – had been born. In the spirit of Descartes, Kant and Laplace considered galaxies to be the result of an evolving universe.

The nineteenth century: giant telescopes and spectroscopy

Herschel and Rosse set new instrumental standards. Herschel's monumental observing programme enlarged the list of nebulae to several thousand, and Rosse discovered the class of spiral nebulae. But their nature remained enigmatic, and it was not certain whether the concept of island universes could survive. When, in the middle of the century, Huggins' spectroscopy added a new dimension to nebular research, it helped to separate them into qualitatively different classes. The advent of photography brought another highly valuable tool. It considerably reduced the personal bias in reporting on their characteristics. However, the real battle about nebulae being island universes was fought in the early twentieth century.

The birth of modern cosmology

The drawn-out debate on the existence of island universes coincided with Einstein's entry into cosmology. The two endeavours were unrelated. Communication between cosmological theoreticians and observers developed only slowly. The first contacts were about redshifts: they had been observed by Slipher, and theoretically predicted by de Sitter.

Einstein and de Sitter, fathers of modern cosmology

In 1917, Einstein published his seminal paper on the large-scale structure of the Universe. There was no direct link to observations, except for the rather 'obvious' assumption of a homogeneous, static universe. The result was certainly very striking: Einstein told us that we live in a closed universe. Should observers be able to provide a mean mass density, then we could even calculate its 'volume', whatever that might mean. This opened the prospect of the whole Universe becoming accessible to observations.

Things became enigmatic when in the same year de Sitter published an alternative. It was a crazy suggestion: contrary to everybody's daily experience, de Sitter's universe contained no matter. However, it held an irresistible fascination to astronomers: it predicted redshifts in the spectra of distant objects, redshifts that ought to increase with the distance of the emitter. Slipher had indeed found redshifts in spiral nebulae; they exceeded by far any other astronomical redshift measured up to that time. They remained the main observational incentive to further pursue cosmology. But de Sitter had sent the theoreticians up a blind alley and for more than ten years they strained their wits to find a way out.

Einstein as well as de Sitter wanted a static universe. But, in 1922, Friedmann realised that these two authors had hit on two special cases, and that the general isotropic and homogeneous solution of Einstein's fundamental equations represented a dynamic universe, either in contraction or expansion. Friedmann did not know about Slipher, and was under the impression that observations did not yet provide useful clues for cosmology. Friedmann's contribution remained unnoticed.

Lemaître discovers the expanding universe

Lemaître found the weak point in de Sitter's approach in 1927. He rediscovered dynamic solutions to Einstein's fundamental equations. Combining his theoretical insight with observations, he found that our Universe was expanding. In the same paper, he discovered what was later named the 'Hubble law' and gave the first determination of the 'Hubble constant'. But Lemaître's paper was published in French and in a Belgian journal that was not widely read. For more than two years no one paid attention to it.

At the time of Friedmann, or shortly afterwards, Weyl and Lanczos reinterpreted de Sitter's solution dynamically. So did Robertson in 1928. However, none of them saw the expanding universe. De Sitter later remarked that once Lemaître had found the solution, it was of such simplicity as to make it appear self-evident, just like Columbus' famous solution of how to stand an egg on its small end.

Observational contributions

What was happening on the observational front? By 1918, Shapley had moved the Sun from the centre to the outskirts of the Milky Way and, by 1925, the battle of the island universes was over: yes, they do exist! Galaxies emerged as building blocks of the Universe. Many observers had contributed to this result, but the decisive piece of evidence – the one that signalled the end of the debate – came from Hubble in an award-winning paper, read on 1 January 1925 at the American Astronomical Society meeting in Washington. Hubble had succeeded in identifying Cepheid variables in three nebulae, Andromeda amongst them. With the Cepheid period–luminosity relationship he could place these nebulae at distances of one million light years, thus locating them far outside the confines of the Milky Way. Hubble did not rest on his laurels. In 1926 he published a long list of apparent nebular magnitudes, and added a distance–magnitude relation.

Inspired by de Sitter's prediction, several observers had, since the early 1920s, tried to find a relationship between nebular distances and redshifts, but without success. Slipher's redshifts measured at the Lowell Observatory were fine, but the distances were very uncertain and of heterogeneous quality. Hubble's paper of 1926 changed the situation by providing a consistent set of distances. The following year, Lemaître was the first to profit from them. Stimulated by his theoretical knowledge, he calculated the rate of expansion: the Hubble constant.

Nebular redshifts also began to be measured at Mount Wilson; Humason became the acknowledged expert. In 1929, Hubble did what Lemaître had done in 1927: he compared distances and redshifts of practically the same set of nebulae. However, by then, the grip on distances had tightened and Humason had determined some new redshifts, one of them going far beyond Slipher's nebulae. Hubble's observationally found linear redshift–distance relationship of 1929 established a new landmark, which was further consolidated by Humason and Hubble in 1931.

Eddington and de Sitter spread the gospel

Hubble's finding much impressed de Sitter, who did his own follow-up redshift–distance determination. In January 1930 he told Eddington and his group about that observational milestone. They had no theoretical solution ready to explain Hubble and de Sitter's results. When Lemaître learnt about their deadlock, he sent them his solution of 1927 – and the scales fell from their eyes. The concept of an expanding universe was soon widely accepted, although dissenting voices were heard and it was debated for another thirty years.

Einstein had dismissed Friedmann as well as Lemaître. But, in the spring of 1931, he withdrew his former rejections. One year later he teamed up with de Sitter for a short publication, where they gave their favoured cosmology. They discarded the cosmological constant, Λ, and opted for a spatially flat, ever expanding universe. It became known as the Einstein–de Sitter universe and was for a long time the standard cosmological model.

The seed for the Big Bang, and the first reference to vacuum energy

Even before Einstein's conversion, Eddington and Lemaître had begun investigating the early phase of the expansion. Eddington favoured a universe with an infinite past that had spent most of its time in Einstein's quasi-static state, before finally entering into expansion. Lemaître proposed what later became known as the Big Bang. He saw the highly condensed initial state as either a real beginning, or a transition between two successive universes that, however, did not communicate with each other. They both judged the cosmological constant to be an important element of cosmology, carrying the main responsibility for expansion.

In 1933, Lemaître equated the cosmological constant, Λ, with vacuum energy. That idea had to wait a long time before it became fashionable.

How different is today's cosmology?

How has cosmology changed since the early 1930s, when the expanding universe was accepted as the new paradigm? Which astronomical pictures should we show de Sitter, Eddington, Einstein, Friedmann, Hubble and Lemaître to impress them with the most striking novelties?

There are four major advances and discoveries that immediately spring to mind: the agreement of the timescales of stellar ages and the 'expansion age', dark matter, the cosmic microwave background radiation (CMB) and dark energy. However, additional, though less sensational new developments in astrophysics have equally enriched cosmology. The discovery of stellar nucleosynthesis and the distinction between primordial and stellar origin of the elements are such landmarks. And our much-increased knowledge about galaxies, their constitution and their histories, and the web-like distribution of matter in the Universe would not fail to impress them.

Whereas the observational sets of Lemaître (1927) and Hubble (1929a) were restricted to galaxies with redshifts, z, of less than approximately 0.01, corresponding to a look-back time of less than 130 million years, the Hubble Ultra Deep Field observations of 2003/4 surveyed the Universe out to $z \approx 6.5$, or close to 13 000 million years in the past. The enormous progress in instrumentation is a decisive factor in that progress. At least as important as the improvements in the observation of visible light is its extension into the wavelength domain of the ultraviolet, X-rays and gamma-rays, as well as infrared and radio, in short: the opening up of the spectral range.

Fig. 18.1 Telescopes 1908 and 2008. Left: The Mount Wilson 60-inch Hale telescope. Commissioned in 1908, it initiated the modern era of large telescopes. (Photograph courtesy of California Institute of Technology; Observatories of the Carnegie Institution, Washington, Mount Wilson Observatory.) Right: The ESO (European Southern Observatories) Very Large Telescope Interferometer (VLTI) located on top of the 2600-m Cerro Paranal mountain in Chile. It consists of four 8-m telescopes and four moveable 1.8-m auxiliary telescopes. (Photograph courtesy of ESO.)

The Hubble constant revisited

The large Hubble constant and the resulting short lifetime of the Universe had greatly bothered the early cosmologists. With $H = 500$ (km/s)/Mpc, the Einstein–de Sitter universe had by now attained an age of $\approx 1.3 \times 10^9$ years. The difference with the much older Earth was only solved in the 1950s. Crucial contributions came in 1956 from Baade, who recalibrated the Cepheid distances, and in 1958 from Sandage, who found that knots identified by Hubble as the brightest stars in more distant galaxies were really H II regions. These reduced the Hubble constant from $H \approx 500$ to $H \approx 75$ (km/s)/Mpc. The various ups and downs of H have come to a relative standstill around this latter value. References to further contributions in the drawn-out quest for a safely verified Hubble constant are given in Van den Berg (1996).

Elemental abundances in the Universe

In the late 1920s, an entirely new picture emerged about the chemical composition of the Universe. The earlier view that the composition of the Sun and stars were similar to the Earth was replaced by the insight that the majority of atoms in the stars are hydrogen, and the heavier elements are a small minority. Further progress in the 1950s slowly led to proportions of approximately 92% hydrogen, 8% helium and 0.1% heavier elements (by number).

Fig. 18.2 Looking back through the ages. Each spot is a galaxy. The picture shows a very small sector of the sky, approximately one tenth the diameter of the Moon. Whereas the nearest galaxies to the Milky Way, e.g. Andromeda, have distances of approximately one megaparsec, corresponding to a look-back time of a few million years, the light of the closest well-defined galaxies in this picture has been on its way for approximately one giga-year (a thousand million years). In the most remote galaxies of this picture we see the Universe when it was 800 million years old, thus, approximately 12 giga-years in the past. The picture was taken between 24 September 2003 and 16 January 2004; the total exposure time was 11.3 days. (Hubble Ultra Deep Field.)

In 1957, Burbidge, Burbidge, Fowler and Hoyle, in their renowned B^{2}FH paper, investigated the creation of the elements, and showed how the process of nucleosynthesis in stars could explain the abundances of essentially all but the lightest elements (Burbidge *et al.* 1957). Seven years later, Hoyle and Tayler attacked the mystery of the cosmic helium abundance (Hoyle and Tayler 1964).

The general He/H abundance ratio is too high to be due to $H \rightarrow He$ fusion in ordinary stars. Hoyle and Tayler (1964) looked at the physical conditions that could produce the required amount of helium. A large number of very massive stars could have delivered the goods; however, they saw no astronomical evidence for such events. They then shifted their attention to a 'radiation origin' of the Universe, where the initial conditions were hotter than several times 10^{10} K. In their calculations the subsequent expansion and drop of temperature resulted in a ratio of He/H ≈ 0.14. Thus, if a stellar origin had to be excluded, the observed He/H ratio suggested that 'the Universe had a singular origin or is oscillatory'. They added that their approximate calculations must be repeated more accurately.

This was done by Peebles. In 1966, after the hot early Universe had been established, based on observation, he drew an observationally, broadly supported picture of how the elements had been created: hydrogen and the major part of helium originated from the Big Bang, the heavier elements were and still are created in stars (Peebles 1966).

Dark matter

An entirely new aspect entered astrophysics with the concept of dark matter. This elusive substance cannot be directly observed in any of our observatories, which are built for detecting electromagnetic radiation. Its presence is inferred from the gravitational effect on light and ordinary (baryonic) matter, and from its effect on the cosmic microwave background radiation.

Dark matter was strongly suspected for the first time in 1933, when Zwicky applied the virial theorem to the Coma cluster in order to obtain its total mass (Zwicky 1933). At that time it did not make the headlines, and Zwicky certainly did not suspect the evasive dark matter of which we talk today. It re-emerged forty years later when, in the 1970s, Vera Rubin hit on it in her analysis of rotational velocities in individual galaxies.

Dark matter is also invoked to explain the lensing of background galaxies caused by intervening massive, compact clusters of galactic matter. The bending of the light from the background sources requires much more gravitation in the lensing clusters than the observable matter can generate. The existence of dark matter is also supported by the analysis of the CMB (see below).

The different pieces of evidence indicate that the Universe contains about five times as much mass in the form of dark matter as that in the form of our well-known matter, the baryonic particles, including electrons. Yet, in spite of all efforts to come to grips with this powerful source of gravitation, we are still ignorant of its nature. For some, however, dark matter is a phantom, indicating that the laws of gravitation need modification.

Fig. 18.3 Dark matter as gravitational lens. The galaxies of Abell 2218 distort images of more distant galaxy populations into arcs. (Hubble Space Telescope Wide Field Planetary Camera 2.)

The cosmic microwave background radiation

The discovery of the cosmic microwave background radiation (CMB) was the most convincing observational support for the Big Bang. It led finally to its general acceptance and ultimately to the demise of the Steady State theory of Bondi, Gold and Hoyle. Central to that story stands the discovery of an 'unaccounted-for antenna temperature' by Penzias and Wilson (1965) and its theoretical interpretation by the astrophysical group of Dicke, Peebles, Roll and Wilkinson (Dicke *et al.* 1965).

Shortly after the Second World War, the cosmic microwave background radiation had been predicted by Gamow as a relic from the extremely hot earliest stage of the Universe (Gamow 1948). In the same article he also volunteered a cosmogonic speculation. According to Gamow, 'one should imagine the original state of matter as a very dense over-heated neutron gas which could have originated (if one lets one's imagination fly beyond any limit) as the result of hypothetical universal collapse preceding the present expansion.' His mental flight 'beyond any limit' led him to a cyclic universe. Gamow might have dreamt up these Big Crunch and Big Bang images himself, or he might have inherited them from Friedmann, who, for a short time, had been his teacher in Leningrad. Lemaître had also speculated on a cyclic universe, and Einstein assumed it in his publication of 1931.

Gamov's extremely hot early Universe has been cooling ever since the Big Bang. Hydrogen, the chief component of the Universe, was completely ionised at those extreme temperatures: there were just photons, neutrinos,

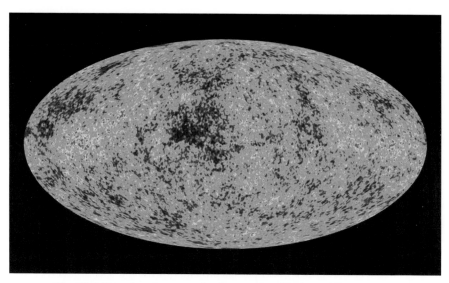

Fig. 18.4 Cosmic microwave background radiation. Differences in intensity show tiny differences in the milli-degree range. The pattern is the background for the present structure of the Universe. (Courtesy of NASA/WMAP Science Team.)

free electrons, and nuclei made of neutrons and protons. Photons are easily scattered by free electrons, this interaction keeps particles and radiation at the same temperature. The expansion of the Universe affects the wavelengths of the photons in the same sense: radiation shifts to longer wavelengths. Therefore, the temperature of the blackbody radiation diminishes. As long as the photons were able to ionise neutral hydrogen, there were plenty of free electrons to guarantee strong interaction between the particles and radiation. When the radiation temperature fell below $\approx 3000\,$K, protons and electrons finally combined to become neutral hydrogen, so there were hardly any free electrons left to scatter radiation: particles and radiation decoupled, and the mean free path of the photons became practically infinite.

The epoch of the transition corresponds to a redshift of $z \approx 1100$ and is thus more than 13 billion years in the past. Photons have ever since travelled practically unaffected by particles. They arrive evenly from all directions from the 'surface of last scattering', which they left when the temperature was approximately 3000 K. This 'surface of last scattering' is a kind of cosmological photosphere. We cannot see beyond that photosphere, but we can deduce quantities from its minute inhomogeneities, some of which must have been fixed before decoupling occurred. The observed inhomogeneities arose from the growth of tiny perturbations, imprinted through several different mechanisms. The power spectrum of the temperature anisotropy, in terms of the angular scale, contains information on these mechanisms and on other properties, such as the density of dark matter.

As long as photons strongly interact with electrons, photon pressure can carry baryons along. Because photons have longer mean free paths than electrons, their diffusion can wipe out baryon inhomogeneities (Silk damping). However, that only works up to a limiting size, approximately given by the mean free path of the photons. The scale below which inhomogeneities will not survive Silk damping shows up in the power spectrum of the CMB. This could imply that mass agglomerations on the scale of galaxies formed before individual stars.

The background radiation also allows us to observationally define an inertial frame of the Universe, with a fundamental observer who is at rest with respect to that inertial frame.

The discovery of the CMB must certainly have come as a profound satisfaction to Lemaître. It vindicated and finally established his pet idea of an expansion out of a near-singularity. The discovery was announced in July 1965; Lemaître died in June 1966.

The structure of the Universe

Einstein assumed a homogeneous universe. Despite occasional question marks, it was generally accepted as a valid assumption. Clusters of galaxies had already been discovered in the 1920s, but they did not jeopardise the overall concept of homogeneity.

However, in the late 1970s, a different kind of structure began to emerge. Gregory and Thompson (1978) showed the string-like presence of a population of galaxies linking the two widely separated clusters Coma and A1367. They also showed a large inter-cluster region devoid of galaxies. Some years later, the same authors published further evidence for super-clustering and voids (Gregory and Thompson 1982). This pattern of a filamentary structure with large voids emerged even more clearly when the sample of galaxies was enlarged, as in the study of Geller and Huchra (1989).

The existence of this cosmic web has since been confirmed, in particular by the 2dF Galaxy Redshift Survey. Clusters of galaxies connected by filaments and sheets of galaxies, surrounding vast and almost empty spaces, show the highly collapsed material content of the Universe on the megaparsec scale. They are another example of Nature's complex hierarchical structure.

The refinement of the CMB measurements has added crucial information about the early phase of the Universe. Together with theoretical studies and the distribution of galaxies seen in the 2dF survey, we may now look at galaxies – the building blocks of the Universe in the 1930s – as foam on the underlying dark matter. The present spatial distribution of dark matter is being calculated as an

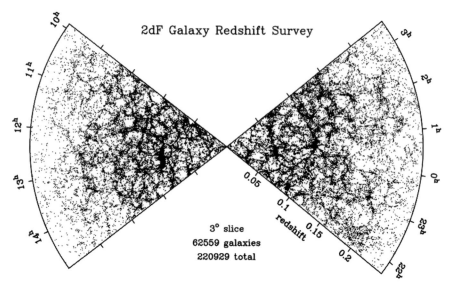

Fig. 18.5 The structure of the Universe. Distribution of galaxies in a slab of about 3 degrees thickness, out to a distance of approximately 1000 Mpc. Each point shows the location of a galaxy. (2df Galaxy Redshift Survey; Colless *et al.* 2001.)

evolutionary result, originating in the earliest inhomogeneities of the inflationary beginning of the Universe. The agreement between observations and the model based on inflationary assumptions is a welcome support for the speculative theory.

Inflation

Does the Universe have a beginning? If it does, what was it? Since the 1980s, inflation has become a central concept in theoretical cosmology. It does not really answer the question whether the Universe has a beginning, but it claims to explain the origin of expansion. In contrast to the speculation of Lemaître, who saw the initial expansion breaking out of a single complex atom that contained all the matter of the Universe, the adepts of inflation see expansion originating in a tremendous burst, powered by some hypothetical field, with its origin in quantum fluctuations much below the atomic level. In a tiniest fraction of a second, the Universe, including of course the small fraction that contains all that we can see today, is blown up to enormous dimensions. During that initial burst, any spatial curvature that may have shaped our now observable volume was stretched out into the now spatially practically flat Universe. Original quantum fluctuations were stretched by the inflationary expansion to structures that eventually evolved into the patterns seen today: large clusters of galaxies, connected by filaments of galaxies surrounding vast voids.

Considerable quantitative observational advances came with the satellites COBE (Cosmic Background Explorer) in the early 1990s and WMAP (Wilkinson Microwave Anisotropy Probe) in the new millennium. Observing the CMB provides information about how the Universe looked at the time when free electrons and protons combined to form stable neutral hydrogen. That epoch, often called by the misnomer 'time of recombination', occurred approximately 380 000 years after the Big Bang.

Accurate measurements of the anisotropies show that the Universe must be nearly flat. On the other hand, we have known for quite a while that the total amount of baryons and dark matter is considerably below the critical density necessary for a spatially flat universe. Hence, an additional energy component must exist.

In 1998, two observational teams showed that faraway supernovae are, on average, fainter than would be expected (Riess *et al.* 1998, Perlmutter *et al.* 1998). This might indicate that the expansion of the Universe is accelerating. Thus, the evidence of a continued action of the cosmological constant has become even more compelling. It now looks as if the expansion rate had initially decreased, but approximately 5×10^9 years ago switched from deceleration to acceleration. The redshift of the galaxies that testify to this change is approximately $z = \Delta \lambda / \lambda \approx 0.46$. This discovery has strengthened belief in inflation and in the fundamental influence of the cosmological term.

But beware, critical voices are warning against assuming Type Ia supernovae to be unchangingly valid standard candles for the whole duration of cosmic history. And there are even more fundamental doubts being sown: they concern our assumption about homogeneity of the spatial fraction of the Universe. Could the growth of structure, with its vast variation in densities, cause spatial fluctuations in the metrics, which might then mimic an accelerating expansion?

In 1933, Lemaître was the first to equate the cosmological constant to vacuum energy (Lemaître 1934). However, this tempting idea seems to have hit on a profound discrepancy. Theoreticians are still unable to calculate the vacuum energy density in quantum field theories. When making order-of-magnitude estimates they arrive at energy densities of $\approx 10^{110} \, \mathrm{erg/cm^3}$, whereas the present value of the cosmological constant corresponds to an energy density of $\approx 10^{-10} \, \mathrm{erg/cm^3}$. This discrepancy of a factor $\approx 10^{120}$ is known as the 'cosmological constant problem' (see Straumann 2007).

The concordance model

It is widely agreed that at the beginning of the twenty-first century the concordance model is the best approximation to our observational and theoretical knowledge. Because it contains a fair share of speculation, no one should be surprised if some of its assumptions or doctrinal beliefs might at a future date have to be revised. The cosmological 'concordance' model agrees that our Universe began in a Big Bang, approximately 13 to 14×10^9 years ago. According to our present knowledge, the energy content of the Universe is composed of approximately 5% ordinary matter, 25 % dark matter and 70% dark energy. Thus, approximately two thirds of the energy in the Cosmos is associated with the cosmological constant Λ, or something similar. The Universe is spatially flat, but the 4-dimensional spacetime is curved. The present expansion of the Universe began in a tremendous inflationary burst within a tiny split of the very first second. Between 10^{-35} and 10^{-33} seconds after the beginning, the Universe expanded by a factor 10^{30}. Inflation does not offer any explanation for the time before that 'beginning'. As mentioned at the beginning of this section, at present the concordance model delivers the best approximation to observations; this is no guarantee for its validity.

Annihilation of particles and antiparticles – first protons and antiprotons, later electrons and positrons – produced the radiation field, which eventually became our observed CMB. The minimal surplus of particles over antiparticles, whose origin is still unknown, is the baryonic matter of stars, galaxies, intergalactic gas, stars, the Moon, the Earth, human beings – and the vintage port, which we will presently need.

A glass of port

If we had Einstein, de Sitter, Eddington, Friedmann, Lemaître and Hubble gathered around a table for a glass of port, how would they comment on present-day cosmology?

Dark matter would probably be the biggest surprise to all of them. They would be thrilled by the cosmic microwave background radiation with its tiny inhomogeneities, that messenger from the beginning of the Universe. They would certainly spend some time discussing black holes. They are not just mathematical curiosities, they do exist; a massive one sits in the very centre of our Milky Way. Would they join the speculation on multiple universes? And, how would they judge inflation? Lemaître might remind them that he had speculated about the primeval atom; why should the next generation not cherish its own pet? – which might well be superseded in its turn. They would certainly be intrigued when confronted with the observed three-dimensional honeycomb structure of the visible Universe, and the possible connection to quantum fluctuations in the earliest Universe.

When talking about their personal feelings, Hubble would, no doubt, be pleased that his name was attached to the first large telescope that has circled the Earth far above ground. He would marvel at the breathtaking multitudes and sizes of observational facilities, with the opening of the spectral range from γ- and X-rays through to the ultraviolet, past the visual, the infrared, and into the radio domain. Maybe he would regret, with some nostalgia, those days when the astronomer, to whom observing time had been allocated, sat close to the focal point of the telescope when the observation was taken. Einstein and de Sitter would hardly be upset by the return of the cosmological constant; would they want to join the speculations on inflation? Eddington and Lemaître would see themselves vindicated for clinging stubbornly to the cosmological constant. Eddington might even become reconciled to Lemaître's Big Bang, seeing that it has not been turned into a devious stratagem for smuggling religion into science. Lemaître with his expanding universe of 1927, the Big Bang, and his early association of the cosmological constant with vacuum energy, would be three times winner, and could thus not avoid footing the bill for the second bottle. No hesitation about the wine: Vintage Port 1927 – the Douro valley produced outstanding wines and most shippers declared it a vintage. The illustrious companions would certainly lift their glasses in honour of Friedmann, and Einstein could do penance by announcing the toast.

They would have good reasons to be well satisfied after seeing their unrelenting efforts crowned with success, cosmology thriving, and their seminal labour being vigorously pursued.

Mathematical Appendix

The general theory of relativity, published by Einstein in 1915, was a scientific revolution (Einstein 1915a,b, 1916). The formerly absolute space, uninfluenced by its content, went through a metamorphosis. Its matter now structures space – matter being understood as anything to which energy can be attributed. The new concept of space, time and matter also called for new mathematics. Riemannian geometry replaced Euclidean geometry, which fits our daily experience so well, and tensors replaced the scalar gravitational potential. Now, space, time and gravitation are unified into a four-dimensional curved spacetime.

This appendix is not meant to be an introduction to General Relativity. Instead, it gives some mathematical comments on the early key discussions on the expanding universe, which should be read in parallel with the corresponding sections of the main text. The following sections carry the same titles as the corresponding sections of the main text, and some duplication has been allowed to facilitate reading.

1. Chapter 6: The early cosmology of Einstein and de Sitter

Some fundamental relations

We outline here some of the concepts that were employed by those involved in the cosmological debate of the 1920s and early 1930s. We present them from today's point of view, although some parts of the theory, especially its mathematical formulation, had not been developed as it is today. However, when discussing the original texts, we stay as close as we can to the original notation.

Basic principles

General Relativity is based on and at the same time extends Newton's theory of gravitation, as well as his equations of motion.

1. Space and time form a four-dimensional manifold, called spacetime. Its geometrical properties are described by its metric, g_{ij}. The metric is determined by gravitation, and

the components are calculated from Einstein's fundamental equations. The metric may depend on the location in space and time.

2. The physical laws are the same in every coordinate system; mathematically speaking, they have to be generally covariant.

3. Locally, the gravitational field can be (almost) transformed away by choosing a coordinate system in which the metric g is flat. In such 'local inertial systems' the laws of Special Relativity are valid.

The interval between two events occurring at x^i and $x^i + dx^i$ is given by ds, where $ds^2 = g_{ij} dx^i dx^j$, and the indices i, j run from 1 to 4.

A general spacetime has no symmetries and, when looking in different directions, things might appear differently. It is not evident a priori that the general spacetime can be split into space and time. However, astronomical observations suggest that we live in a homogeneous and isotropic space (not spacetime!). Choosing a 'natural' coordinate system helps when visualising models. We try to choose coordinates that mirror observed qualities. The model should therefore allow for splitting the 4-dimensional spacetime into a 3-dimensional homogeneous and isotropic space and a 1-dimensional time. Thus, it becomes possible to foliate spacetime into space-like slices by a time function, such that each slice is homogeneous and isotropic. Such a space-like slice can itself be curved or flat.

Einstein's field equations relate the geometric structure of spacetime to its material content. Even today, they cannot be solved in full generality; assumptions have to be specified that are relevant for solving particular problems, e.g. exploring the environment of black holes, lensing by galaxies, or the evolution of the Universe. When trying to solve cosmological problems, such as finding the structure and dynamics of the Universe, the symmetries suggested by observations help to simplify the mathematical task.

It is important to keep in mind for the concept of curved space that this is not a curvature of a three-dimensional object into a fourth dimension. The fourth dimension in General Relativity is time. The curvature of the three-dimensional space concerns the inner structure of that space. For a discussion of n-dimensional spaces and curvature, see the same section of the main text.

In classical mechanics, space and time exist on their own, with no relation to the presence or absence of particles and fields. The solution of the equation of motion describes how a particle is moving in the presence of an acting force. General Relativity differs because it considers gravitation as structuring spacetime; Einstein's field equations describe the structure. The equation of motion describes the trajectory of a particle in that space shaped by gravitation. For a particle, on which only gravitation is acting (freely falling), the resulting trajectory is a geodesic. A geodesic line traces the shortest (or longest) distance between two events. In Euclidean space, a geodesic is a straight line. However, in our cosmological environment, structured according to General Relativity, a geodesic is generally no longer a straight line. Thus, in spherical space it is a great circle. A geodesic can be seen as a 'generalisation' of a straight line. We may choose an inertial coordinate system in such a way that the coordinate lines are geodesics.

In the classical picture of the Universe, space and time are kinds of reality 'a priori', whereas the material bodies are a kind of secondary reality. General Relativity avoids that split into two realities. The matter it contains structures space, shaping space in a non-Euclidean way. Matter has to be understood in a broad sense, including particles as well as radiation.

The field equations

In General Relativity, physical laws are given in the form of Einstein's field equations:

$$R_{ij} - \frac{1}{2}g_{ij}R = 8\pi G T_{ij} \tag{1.1a}$$

or

$$G_{ij} = 8\pi G T_{ij}. \tag{1.1b}$$

The speed of light has been set to $c = 1$. In Eq. (1.1b), the field equations are expressed with the help of the Einstein tensor, defined as

$$G_{ij} = R_{ij} - \frac{1}{2}g_{ij}R, \tag{1.2}$$

where R_{ij} is the Ricci curvature tensor, R the scalar curvature tensor, g_{ij} the metric tensor, G is the gravitational constant and T_{ij} denotes the energy–momentum tensor. The indices i, j run from 1 to 4, and, if we restrict our cosmological models to a homogeneous, isotropic universe, the indices 1, 2, 3 refer to the spatial coordinates, and 4 refers to time.

The set of Einstein's field equations is a tensor equation relating a set of symmetric 4×4 tensors. Due to the symmetry, they reduce to 10 non-linear, second-order, hyperbolic partial differential equations for the metric tensor field g_{ij} with its 10 unknown components. As a result of geometry, the left-hand side of Einstein's equations fulfils the so-called Bianchi identity

$$\nabla^i G_{ij} = 0. \tag{1.3}$$

This gives 4 constraints and reduces the number of independent equations to 6. It allows a freedom of choice for the 4 spacetime coordinates. Through the Bianchi identity, the Einstein equations imply

$$\nabla^i T_{ij} = 0. \tag{1.4}$$

This was often thought to express energy conservation. However, as we shall see later, General Relativity does not know a general law of energy conservation. For a perfect fluid, Eq. (1.4) contains all the information about the motion of matter.

For our cosmological models, the energy–momentum tensor, which consists of 4×4 terms, can be much simplified. It is assumed that each spatial region evolves independently, and does not act in a time-dependent way on other regions. This eliminates the components $T_{i4}, i = 1, 2, 3$. Thus, $T_{i4} = T_{4i} = 0$. The off-diagonal elements T_{ij} ($i, j = 1, 2, 3$) contain information on viscosity. Such effects are neglected. The energy–momentum tensor is therefore reduced to the diagonal elements

$$T_{11} = T_{22} = T_{33} = p, \quad T_{44} = -\rho, \tag{1.5a}$$

or

$$T_{jj} = (p, p, p, -\rho) \tag{1.5b}$$

where p stands for the pressure, and ρ for the energy density. The pressure could be due to gas or radiation. The energy density contains, for example, the rest mass and radiation. Different authors employ different units and conventions on the sign of these components, but the principle is always the same: the Einstein equations link the curvature of spacetime, G_{ij}, to its energy and momentum content, T_{ij}. Thus, mass and energy enter the curvature of spacetime. The solutions to Einstein's equations are the metrics g_{ij} of spacetimes; they describe the gravitational fields.

How does one arrive at Einstein's field equations? A reasonable theory of gravitation should contain the Newtonian theory as a limit because it has shown its validity in our everyday experience, as well as in celestial mechanics. One may therefore start with the Newtonian theory. We have to keep in mind that the set of equations (1.1) are field equations, where the g_{ij} describe the gravitational field. In his cosmological paper of 1917, Einstein gave essentially the following heuristic considerations. In Newton's theory, gravitation is described by a potential Φ. Its negative gradient is the acceleration imparted to a test particle. The Poisson equation allows us to calculate the spatial distribution of the potential as a function of ρ, the density of matter:

$$\Delta \Phi = 4\pi G \rho. \tag{1.6}$$

The left term contains the second derivatives of the Newtonian potential; the right term contains ρ, the density of matter. The Poisson equation follows from Newton's inverse-square law of gravitation. We now have to find a set of equations that in General Relativity replaces the Newtonian potential equations. As the geometry of spacetime absorbs the effect of gravitation, we have to relate the metric of spacetime to the distribution of matter.

The structure of spacetime is defined by its metric, g. This structure is such that a freely falling particle follows a geodesic trajectory. Investigation of motion involves time, thus g_{44}. This geometrical representation of gravitation has superseded the Newtonian concept of gravitation. In the Newtonian limit for a weak gravitational field, with the Newtonian potential Φ satisfying the potential equation (1.6), and choosing appropriate coordinates, we obtain for the g_{44} component of the metric:

$$g_{44} \approx -(1 + 2\Phi). \tag{1.7}$$

For a vanishing field, $\Phi \to 0$, and we have the classical Minkowskian metric.

For a test particle moving in a quasi-stationary weak gravitational field, we find that there exist nearly Lorentzian coordinate systems, and we can write the metric g as

$$g_{ij} = \eta_{ij} + h_{ij} \text{ with } |h_{ij}| \ll 1, \tag{1.8}$$

and η stands for the metric in flat spacetime. Deviations from Minkowskian geometry increase with increasing density ρ.

The equation of motion

How do particles move in that field? We have to replace the Newtonian equation of motion

$$\frac{d^2 x}{dt^2} = -\nabla \Phi. \tag{1.9}$$

We replace Φ by $h_{44} \approx -2\Phi + C$, C being a constant. Neglecting the small terms for a quasi-stationary field, we obtain the approximation

$$\frac{d^2x^j}{dt^2} \approx \frac{1}{2}\frac{\partial}{\partial x^j}h_{44}. \tag{1.10}$$

For an isolated system the potential Φ should go to zero at infinity. This applies to h_{44} as well, and we therefore find $g_{44} \approx -(1 + 2\Phi)$, as given in Eq. (1.7).

We now turn to the Newtonian potential of Eq. (1.6). $\Delta\Phi$ will be taken over by the corresponding curvature term of General Relativity, and the force term on the right by the energy–momentum tensor

$$T_{ij}u^iu^j \leftrightarrow 4\pi G\rho. \tag{1.11}$$

This leads to the General Relativity equivalent of the Poisson equation (1.6):

$$R_{ij} = 8\pi G(T_{ij} - \frac{1}{2}g_{ij}T). \tag{1.12}$$

We now compare trajectories of particles and light, calculated in the classical Newtonian way, with the formalism of General Relativity. In the absence of any external forces the particle moves on straight lines in Newton's classical space, as well as in the Minkowskian space of Special Relativity. For the Minkowski space of Special Relativity, we write for the line element

$$ds^2 = dx_1^2 + dx_2^2 + dx_3^2 - dt^2. \tag{1.13}$$

The line element ds is the distance between two infinitesimally close points in spacetime; its size does not depend on the coordinates employed for measuring. Should gravitation be active, this has to be specified explicitly. The presence or absence of gravitation has no influence, neither on Newton's classical Euclidean nor on Minkowski's structure. Whereas the theory of special relativity unifies space and time, the theory of General Relativity unifies space, time and gravitation.

General Relativity differs. The structure of spacetime is determined by gravitation, which is transformed into a geometrical curvature. This effect is included in Γ, the Christoffel symbol of curvature. The equation of motion in General Relativity is

$$\frac{d^2x^i}{ds^2} + \Gamma^i_{kl}\frac{dx^k}{ds}\frac{dx^l}{ds} = 0. \tag{1.14}$$

It determines how particles move under the influence of a gravitational field. But it does not tell us about the metric. The metric is subject to the field equations linking the geometry of spacetime to the energy–momentum tensor, thus to the mass–energy distribution. Γ is defined by the metric tensor

$$\Gamma^i_{kl} = \frac{1}{2}g^{im}\left(\frac{\partial g_{mk}}{\partial x^l} + \frac{\partial g_{ml}}{\partial x^k} - \frac{\partial g_{kl}}{\partial x^m}\right), \tag{1.15}$$

where $g^{im} = (g^{-1})^{im}$ denotes the components of the inverse of g_{im}.

From the equation of motion, $\frac{d^2x^i}{ds^2} = -\Gamma^i_{kl}\frac{dx^k}{ds}\frac{dx^l}{ds}$, we see that in General Relativity, $-\Gamma^i_{kl}\frac{dx^k}{ds}\frac{dx^l}{ds}$ corresponds to the gravitational force. It describes how the gravitational field acts on an otherwise 'free particle': the particle is constantly accelerated.

Differential geometry delivers a connection between the metric tensor, g, the Christoffel symbol, Γ, and the Riemannian curvature tensor, R.

g: the metric tensor corresponds to the gravitational potential.

Γ: the Christoffel symbol consists of derivatives of g; it represents the gravitational field, and Γ^i_{kl} are its components.

R: the Riemannian curvature tensor is a property of spacetime and does not depend on the coordinate system. It can be expressed through Γ and derivatives of Γ. It gives the variation of the gravitational field. There is a qualitative difference between Γ and R. Γ is defined in local coordinates and can be locally transformed away. This leads to a locally Minkowskian structure. The curvature tensor R is a geometrical property of space; it cannot be transformed away. The Riemannian curvature tensor can be expressed in local coordinates through Γ:

$$R^l_{ijk} = \Gamma^p_{ik}\Gamma^l_{jp} - \Gamma^p_{jk}\Gamma^l_{ip} + \Gamma^l_{ik,j} - \Gamma^l_{jk,i}. \tag{1.16}$$

R and g are tensors representing global properties of spacetime. They do not depend on the particular choice of coordinates. Γ is not a tensor; it is defined by the local coordinate system.

To show the physical significance, we recall Einstein's example of the observer in the windowless free-falling lift. Suppose the free fall towards a central point-mass passes a point at distance d. R(d) and g(d) are determined by d and the mass at the centre. Their values do not depend on the coordinates chosen to calculate them. The free-falling observer feels no acceleration, and from the equation of motion (1.14) it is clear that there must be a coordinate system in which the acceleration vanishes, which implies Γ = 0. In contrast, we now look at an observer who stands on a platform fixed at the same distance d from the point-mass. R(d) and g(d) are the same as before. However, the platform prevents the free fall towards the centre, and acts as an accelerating force on the observer who is now 'at rest'.

The static universe of Einstein

Einstein (1917) starts his cosmological investigation with classical deliberations. If the Maxwell–Boltzmann distribution for a gas is applied to the stars, then a Newtonian stellar system cannot exist. This is because the finite potential difference between the centre and the spatially infinite corresponds to a finite ratio of densities. Therefore, if the density of matter, ρ, vanishes at infinity, it will also vanish in the centre. Einstein then looks at the Poisson equation, $\Delta\Phi = 4\pi G\rho$ (Eq. 1.6), which gives the gravitational potential Φ from the distribution of ρ. A boundary condition has to be added so that the potential reaches a finite value at infinity. Einstein supplements the Poisson equation with a universal constant, Λ,

$$\Delta\Phi - \Lambda\Phi = 4\pi G\rho. \tag{1.17}$$

Then,

$$\Phi = -\frac{4\pi G}{\Lambda}\rho_0 \tag{1.18}$$

is a solution of the extended Poisson equation, ρ_0 being the mean density in the Universe. That would provide a spatially homogeneous universe of infinite extent. Yet, there is no physical justification for modifying the classical Poisson equation in such an arbitrary manner.

Einstein applies a similar reasoning to General Relativity in cosmology. The Poisson equation is replaced by Einstein's field equation for gravitation,

$$G_{ij} = -\kappa(T_{ij} - \frac{1}{2}g_{ij}T), \tag{1.19}$$

where $\kappa = 4\pi G$ is the Einstein constant. To find a cosmological solution, boundary conditions at infinity have to be added. Einstein describes in the 1917 publication his tortuous way to solving the boundary problem. He wants a static universe, but realises that his original field equations do not give such worlds. The introduction of the cosmological constant Λ (Einstein calls it λ) gives him a spatially closed universe. Thus further boundary conditions become superfluous. Or, as Eddington said, Einstein solved the boundary problem by abolishing infinity. Formally, there was no problem in modifying the field equations by adding the universal constant Λ to obtain

$$G_{ij} - \Lambda g_{ij} = -\kappa(T_{ij} - \frac{1}{2}g_{ij}T). \tag{1.20}$$

Einstein emphasises that the known laws of gravitation do not justify the introduction of Λ; its inclusion is motivated by the quest for a static universe. It does not worry him that Λ leads to a positive curvature of space. That is no argument against Λ, he says, because the presence of matter will anyway result in a curved space.

Einstein added Λ to the left-hand side of the fundamental equation, which contains the information on geometry. It could equally well be added to the right-hand side, where matter and energy contents are given. In the first instance, one might interpret Λ as a property of space, in the second, as an additional energy term.

Einstein's universe is spatially homogeneous as well as isotropic. He naturally splits his 4-dimensional spacetime into 3-dimensional space and 1-dimensional time. This shows up in the line element

$$ds^2 = g_{ij}dx^i dx^j. \tag{1.21}$$

As his world is static, all g_{ij}, $i, j \leq 3$ have to be independent of time, and he also has $g_{14} = g_{24} = g_{34} = 0$, and $g_{44} = 1$. Einstein then argues that an even distribution of matter will result in a constant curvature. For a given time x_4, the closed continuum of the x_1, x_2, x_3 will therefore be a spherical space. For that space, he finds the following relation between the cosmological constant, Λ, the mean density, ρ, and the radius of curvature, R:

$$\Lambda = \frac{\kappa\rho}{2} = \frac{1}{R^2}. \tag{1.22}$$

This relation allows him to calculate the size and the total mass of the Universe from the observationally determinable density:

$$M = \rho 2\pi^2 R^3. \tag{1.23}$$

In 1926, Hubble employed Eq. (1.22), transformed into cgs units and parsecs, to derive from his observationally determined mean density of $\rho = 1.5 \times 10^{-31}$ g/cm^3, the radius of the Universe:

$$R = \frac{c}{\sqrt{4\pi G}} \frac{1}{\sqrt{\rho}} = 2.7 \times 10^{10} \text{ parsec}. \tag{1.24}$$

Einstein also emphasised that his extended version of the field equations conserves energy and momentum. However, as already mentioned, the concept of energy and momentum conservation will be questioned by Weyl in 1921, when he pointed out that in General

Relativity there is no general conservation law for energy and angular momentum, nor can these quantities, in general, be properly defined (Weyl 1921, p. 246).

The static universe of de Sitter

The outstanding characteristic of de Sitter's universe was that it was devoid of matter, and its outstanding phenomenon was the 'de Sitter effect', a slowing down of time with increasing distance from the observer (de Sitter 1917). In de Sitter's model, the line element has the form

$$ds^2 = R^2(-d\chi^2 - \sin^2\chi(d\theta^2 + \sin^2\theta d\phi^2) + \cos^2\chi dt^2), \tag{1.25}$$

where $\chi = r/R$. For the same angles θ and ϕ for observer and observed, r is the proper radial distance between them, whereas R is the constant curvature, the 'radius' of the Universe. De Sitter calls it system B; system A being that of Einstein. The spatial coordinates do not depend on time. In fact, this universe is static. In de Sitter's model, Λ and R are related by

$$\Lambda = 3/R^2. \tag{1.26}$$

We now come to a crucial property of the original de Sitter model, the *de Sitter effect*. Only on the last three of his 28 pages does de Sitter draw attention to redshifts: 'In the system B we have $g_{44} = \cos^2\chi$. Consequently the frequency of light-vibrations diminishes with increasing distance from the origin of coordinates. The lines in the spectra of very distant stars or nebulae must therefore be systematically displaced towards the red, giving rise to a spurious positive radial velocity.' De Sitter did not derive formulae for the redshift; Eddington did that. For a clock at rest, where χ, θ and ϕ remain constant, he obtained from the line element (1.25)

$$ds = R\cos\chi dt. \tag{1.27}$$

The slowing down of time with increasing distance r also implies that the time for a given cycle, such as an atomic clockbeat, is growing proportionally to $\sec\chi$ (Eddington 1924, section 67). (We recall the notation $\sec(\chi) = 1/\cos(\chi)$.) This effect is also seen in a lowering of the frequency and lengthening of the wavelength of light emitted at distance r. The ratio of received and emitted wavelengths increases according to $(\lambda + \Delta\lambda)/\lambda = \sec\chi \approx 1 + (r/R)^2/2$, or

$$\frac{\Delta\lambda}{\lambda} \approx \frac{1}{2}\left(\frac{r}{R}\right)^2, \text{ for } r \ll R. \tag{1.28}$$

This mechanism was usually referred to as the de Sitter effect. It had been explicitly pointed out by de Sitter that this effect might be responsible for the large redshifts seen in some nebulae. With r increasing further, and thus χ approaching $\pi/2$, the approximation $r \ll R$ is no longer valid. The length of an atomic clock beat further increases according to $\sec(\chi)$. Spectral lines emanating from distant sources will be displaced further to the red, finally disappearing at the horizon for $\chi = \pi/2$. At the horizon, a finite interval ds corresponds to an infinite dt; time has come to a standstill.

In de Sitter's model, there is a second effect leading to a redshift. On page 17 of his 1917 paper he showed that, due to his space-dependent g_{44}, a material particle under the unique influence of inertia does not describe a straight line of constant velocity, but is accelerated towards greater distances. For a particle at rest, $dr/ds = 0$, Eddington (1924, equation 70.22) calculated the acceleration as

$$\frac{d^2 r}{ds^2} = \frac{1}{3} \Lambda r. \tag{1.29}$$

Thus, a particle does not remain at rest unless it is at the origin; for others, there is a tendency to scatter. The expected scattering velocity was cited as an additional explanation for the redshifts. However, de Sitter's universe was meant to be static. Later, Lemaître will point out that de Sitter's choice of coordinates invalidated his explanation of the observed redshifts (Lemaître 1927).

De Sitter's Trojan horse

Interpreting de Sitter's model and redshifts was a difficult task. Many papers were written on the subject. We briefly discuss two of them here. Lanczos found the way out of de Sitter's coordinate labyrinth, but did not recognise the physical significance. Weyl, although he talks about redshifts and observations, remained in a formalistic world. Useful observations were anyway not available.

Lanczos and Weyl worked with a flat 5-dimensional pseudo-Euclidean (Minkowski) space into which they placed a 4-dimensional curved universe; geometrically it is a hyperboloid:

$$x_1^2 + x_2^2 + x_3^2 + x_4^2 - x_5^2 = 1. \tag{1.30}$$

Lanczos' choice of coordinates for the 4-dimensional spacetime results in a metric of the form

$$ds^2 = -dt^2 + \frac{(e^t + e^{-t})^2}{4} (d\varphi^2 + \cos^2 \varphi d\psi^2 + \cos^2 \varphi \cos^2 \psi d\chi^2). \tag{1.31}$$

He set t as time, and φ, ψ, χ as spatial polar coordinates (Lanczos 1922). This gave a spatially closed universe. The radius of curvature varies in time. It grows for positive and negative times to infinity, but cannot diminish. Lanczos had the key to an expanding universe in his hands, but he did not unlock the door. Two years later he returned to cosmology with an article 'About a stationary cosmology in the sense of Einstein's gravitational theory' (Lanczos 1924).

Through Eddington's book on relativity, first published in 1923, Weyl had become aware of nebular redshifts. He wanted to find an explanation and therefore turned to de Sitter's model. There he found that from a star A, an observer B obtains a spectrum, which is shifted to the red according to

$$\frac{\Delta\lambda}{\lambda} = tg\left(\frac{r}{a}\right), \tag{1.32}$$

where r is the distance of the star A from the observer B, and a is the constant radius of the Universe (Weyl 1923b). Weyl then went his own way and derived redshifts in two different ways. He imbedded the 4-dimensional universe in a 5-dimensional pseudo-Euclidean space with

$$\Omega(x) = x_1^2 + x_2^2 + x_3^2 + x_4^2 - x_5^2, \tag{1.33}$$

and

$$ds^2 = -\Omega(dx). \tag{1.34}$$

The equation

$$\Omega(x) = a^2 \tag{1.35}$$

defines a 4-dimensional hyperboloid. The geodesic lines are cut out by the 2-dimensional planes through the origin of the 5-dimensional space. Weyl defined A and B as belonging to the 'same from the origin connected causal world'. He then explored two ways to the redshift:

(1) The redshift seen by B in the spectrum of A is due to the differences in proper time between A and B. From this follows the redshift

$$\frac{\Delta\lambda}{\lambda} = \rho - 1 = \tan r. \tag{1.36}$$

(2) The redshift is due to the Doppler shift from the radial velocity between A and B, as seen in the system of the observer B from the expression

$$v = \frac{dr}{d\sigma}. \tag{1.37}$$

Weyl invoked a universal 'tendency to flee' (Fliehtendenz) as the cause of the displacement of A relative to B, due to the cosmological term Λ. In this connection he introduced what became known as the 'Weyl postulate': he assumed that all the stars of the causal system diverge from a randomly chosen star. The tracks of the stars then resemble a bunch of 'pencil beams originating in one point'. For the relation between r and σ he found:

$$\tan r = ae^\sigma. \tag{1.38}$$

This leads to a relationship between the velocity, v, the redshift, $\Delta\lambda$, and the distance, r:

$$\frac{v}{c} = \frac{\Delta\lambda}{\lambda} = \frac{dr}{d\sigma} = ae^\sigma \cos^2 r = \sin r \cos r, \tag{1.39}$$

which is valid for $v \ll c$. But this result is in conflict with the redshift (1.36) found above. Weyl acknowledges this fact, but does not further comment on it. For small r, they both give

$$\frac{\Delta\lambda}{\lambda} \sim r. \tag{1.40}$$

Weyl ends his article by referring the reader to Eddington's book on relativity for an astronomical interpretation. In his own *Raum, Zeit, Materie*, Weyl mentions redshifts in Appendix III (Weyl 1923a). He has his doubts: 'One cannot yet claim, that our explanation hits the right solution.' In 1930, Weyl returned to this paper after having become aware of Lemaître and Robertson.

2. Chapter 7: The dynamical universe of Friedmann

Friedmann distinguished two classes of premises. To the first class belong the premises of Einstein and de Sitter, which concern the equations for the gravitational potential, the status and the motion of matter. The second class contains the assumption about the geometrical character of the Universe. Within the second class, Friedmann accepts a 3-dimensional spatial world of constant curvature, but this curvature is allowed to vary with time. The world coordinates can be split into three spatial components x_1, x_2, x_3, and a time coordinate x_4. The line element

$$ds^2 = g_{ik}dx^i dx^k \tag{2.1}$$

can be written in the form

$$ds^2 = R^2 \left(dx_1^2 + \sin^2 x_1 dx_2^2 + \sin^2 x_1 \sin^2 x_2 dx_3^2\right) + M^2 dx_4^2. \tag{2.2}$$

R may now be a function of x_4, it is proportional to the radius of curvature, and M may depend on all four coordinates x_1, x_2, x_3, x_4. (We retain the coordinate notation of Friedmann, except for the cosmological constant, which we designate by Λ instead of λ, as was the custom in the early days of relativistic cosmology.)

Friedmann first discussed static worlds, where R is independent of x_4; he called it the stationary world. He recovered Einstein's cylindrical and de Sitter's spherical worlds by choosing for the function R^2 the constant $-R^2/c^2$, where R is the time-independent constant radius of curvature. For Einstein's world, $M = 1$, whereas in de Sitter's model, $M = \cos x_1$. He points out that there are no other stationary worlds.

Friedmann defined as a 'non stationary world' the case where R depends on x_4, and M depends only on x_4. Without loss of generality, he set $M = 1$, and allowed R to become a time-dependent function: $R = R(x_4)$. This resulted in non-stationary worlds. In Einstein's field equations, Friedmann retained the cosmological term Λ; he stated that it could assume any value, including $\Lambda = 0$:

$$R_{ik} - \frac{1}{2}g_{ik}R + \Lambda g_{ik} = -\kappa T_{ik} \quad i, k = 1, 2, 3, 4. \tag{2.3}$$

Neglecting pressure terms, the time-dependent R is determined by Λ and the density ρ. The energy–momentum tensor, T_{ik}, reduces to

$$T_{44} = c^2 \rho g_{44}, \quad T_{4i} = 0, \quad T_{ik} = 0 \text{ for } i, k = 1, 2, 3. \tag{2.4}$$

This results in the Friedmann equations

$$\frac{\dot{R}^2}{R^2} + \frac{2R\ddot{R}}{R^2} + \frac{c}{R^2} - \Lambda = 0, \tag{2.5}$$

$$\frac{3\dot{R}^2}{R^2} + \frac{3c^2}{R^2} - \Lambda = \kappa c^2 \rho. \tag{2.6}$$

The first equation he obtained by setting $i = k = 1$, 2, 3, and with $i = k = 4$ he obtained the second equation (Friedmann 1922). In the last section of the Appendix we come back to Friedmann's equations.

3.   Chapter 9: Lemaître's discovery of the expanding universe

Doubts about de Sitter's choice of coordinates

In 1925, Lemaître took a first step to correct de Sitter's ill-chosen coordinate system (Lemaître 1925a,b). He looked at de Sitter's line element

$$ds^2 = R^2\left[-d\chi^2 - \sin^2\chi(d\theta^2 + \sin^2\theta d\phi^2) + \cos^2\chi d\tau^2\right], \tag{3.1}$$

where R is the constant radius of the four-dimensional universe with the coordinates χ, θ, ϕ, τ. This is a static universe. The principle of homogeneity implies that in the four-dimensional universe any point is equivalent to any other point. But in de Sitter's assignment of the coordinates this is not the case. The spatial element $R^2(-d\chi^2 - \sin^2\chi(d\theta^2 + \sin^2\theta d\phi^2))$ is indeed a constant and independent of time, and thus delivers a static universe. However, the time element $R^2\cos^2\chi d\tau^2$ in general depends on the spatial position χ, except for the case $\chi=0$. The lines of constant χ, θ, ϕ, which give the direction of time, are only geodesics for the point $\chi=0$ which passes through the origin. This point is therefore singled out and has the properties of a centre. This violates the premises of homogeneity and causes paradoxical results at the horizon, which are due to a coordinate singularity.

Lemaître then demonstrates that the line element may also be written as

$$ds^2 = R^2\left[-e^{\pm 2T}(dx^2 + dy^2 + dz^2) + dT^2\right]. \tag{3.2}$$

To show the fundamental property more clearly, we write this line element as

$$ds^2 = R^2\left(dT^2 - f(T) \times R3\right), \tag{3.3}$$

where $R3$ stands for the 3-dimensional Euclidean space. Due to the time-dependent metric, this model also results in redshifts that increase with the distance of the emitter. But, as reported in the main text, Lemaître was not happy about this Euclidean solution.

The discovery of the expanding universe

In 1927, Lemaître returns to the problem (Lemaître 1927). He now associates the Universe to a rarefied homogeneous gas, where the extragalactic nebulae form the gaseous molecules. For the line element, he chooses the more general expression

$$ds^2 = -R(t)^2 d\sigma^2 + dt^2, \tag{3.4}$$

where σ denotes the spatial volume element, and $R(t)$ stands for the radius of curvature of the three-dimensional space.

Lemaître, as Einstein before him, intends to study a closed universe where energy is conserved. He sets p equal to the radiation density, δ to the energy concentrated in matter, and ρ to the total energy density, thus $\rho=\delta+3p$. Lemaître writes the formula for the 'conservation of energy' as

$$\frac{d\rho}{dt} + \frac{3R'}{R}(\rho + p) = 0. \tag{3.5}$$

With the volume of space

$$V = \pi^2 R^3, \tag{3.6}$$

the formula for energy conservation becomes

$$d(V\rho) + pdV = 0. \tag{3.7}$$

This implies that the variation of the total energy, added to the work done by radiation pressure, is zero. For his application to the present universe, Lemaître then neglects the radiative energy density and sets in his subsequent calculations $\rho = \delta$. However, at the end of the article, he argues that radiative energy must not be dropped altogether because it might cause the initial expansion.

Lemaître's choice of coordinates, as given in Eq. (3.4), respects homogeneity of space. Retaining the image of a terrestrial globe, the meridians are chosen as the time lines, and the circles of latitude as spatial lines. (See the illustrations in the main text.) The world-lines for all points on the same circle of latitude are geodesics; for all of them, time flows in the same way. Thus homogeneity is respected. This is what one could call a 'natural choice of coordinates'. The semi-diameter of the circle of latitude corresponds to the time-dependent curvature of space, $R(t)$. When moving in time along a geodesic world-line, the radius of curvature $R(t)$ increases.

For light emitted at position σ_1 and received at σ_2, we find from (3.4), with $ds = 0$,

$$\sigma_2 - \sigma_1 = \int_{t_1}^{t_2} \frac{dt}{R(t)}. \tag{3.8}$$

Emission and absorption processes occur during the times $t_1 + \delta t_1$ and $t_2 + \delta t_2$, δt_1 and δt_2 being the period of light at emission and absorption. Thus,

$$\frac{\delta t_2}{R_2} - \frac{\delta t_1}{R_1} = 0. \tag{3.9}$$

Wavelength λ and period δt are coupled by the relation $\lambda = c\delta t$. Light emitted with wavelength λ_1 at σ_1 will therefore arrive at σ_2 with wavelength

$$\lambda_2 = \lambda_1 \frac{R_2}{R_1}. \tag{3.10}$$

Subtracting λ_1 from both sides gives

$$\lambda_2 - \lambda_1 = \lambda_1 \left(\frac{R_2}{R_1} - 1 \right), \text{or} \quad \frac{\lambda_2 - \lambda_1}{\lambda_1} = \left(\frac{R_2}{R_1} - 1 \right). \tag{3.11}$$

This relation holds generally.

When a redshift is interpreted as a Doppler shift due to a velocity v, with $v \ll c$, we have

$$\frac{\Delta\lambda}{\lambda} = \frac{v}{c}. \tag{3.12}$$

We enter this expression in relation (3.11), assuming $\lambda_2 - \lambda_1 = \Delta\lambda \ll \lambda_1$, which implies that R has changed little between emission and absorption:

$$\frac{\Delta\lambda}{\lambda_1} = \frac{v}{c} = \left(\frac{R_2}{R_1} - 1 \right). \tag{3.13}$$

Lemaître transforms this expression in order to extract the astronomically determinable quantities v and r, where r is the distance to an extragalactic nebula. He assumes the closed universe to be much larger than the distances to the observed extragalactic nebulae: $r \ll R$. In his own notation, the increase of the radius of the Universe $R(t)$ is given by

$$\frac{\Delta\lambda}{\lambda} = \frac{v}{c} = \frac{R_2 - R_1}{R_1} = \frac{dR}{R} = \frac{\dot{R}}{R} dt = \frac{\dot{R}}{R} r. \tag{3.14}$$

To obtain this result, he replaces in the line element $ds^2 = -R(t)^2 d\sigma^2 + dt^2$, the spatial part $R(t)$ $d\sigma$ by r, and obtains, with $ds = 0$, the relation $r = dt$, thus expressing time in units of r/c, which gives him finally the approximate relation

$$\frac{v}{cr} = \frac{\dot{R}}{R},$$

(3.15)

valid for $r \ll R$. We have in equations (3.14) and (3.15) the first formulation of the *Hubble relation*:

$$v = Hr.$$

(3.16)

H is the Hubble constant, which Lemaître determined from observations.

4.   Chapter 13: Robertson and Tolman join the game

Robertson starts from first principles

In 1928, Robertson studied de Sitter's universe. As others before him, he was fascinated by its prediction that the spectral lines from distant sources would be displaced to the red. But he wanted to replace de Sitter's line element by 'a mathematically equivalent solution which was susceptible of a perhaps simpler interpretation and in which many of the apparent paradoxes inherent in [de Sitter's] solution were eliminated.' (Robertson 1928.) He wrote de Sitter's line element as

$$ds^2 = -\frac{d\rho^2}{1 - \kappa^2 \rho^2} - \rho^2 \left(d\theta^2 + \sin^2 \theta d\phi^2 \right) + \left(1 - \kappa^2 \rho^2 \right) c^2 d\tau^2,$$

(4.1)

where $\kappa = \sqrt{\Lambda/3}$, τ = the proper time of a clock at the origin, and ρ is the distance from the origin. He then applied the coordinate transformation

$$\rho = r e^{\kappa c t}, \quad \tau = t - \frac{1}{2\kappa c} \log\left(1 - \kappa^2 r^2 e^{2\kappa c t} \right),$$

(4.2)

and obtained

$$ds^2 = -e^{2\kappa c t} \left(dx^2 + r^2 d\theta^2 + r^2 \sin^2 \theta d\phi^2 \right) + c^2 dt^2,$$

(4.3)

where $x = r \sin \theta \cos \phi$. This is the line element of a dynamic universe. Like Lemaître one year earlier, he had arrived at a representation that respects homogeneity. But unlike Lemaître, Robertson restricted his investigation to de Sitter's stationary universe and its 'dynamical' interpretation. Robertson says: 'The line element [4.3] is "dynamical" in that its coefficients depend on the time t, whereas [4.1] is not. That the properties of the manifold are, however, independent of t may be seen by introducing a change of scale and time origin by means of the transformations

$$\bar{r} = r e^{\kappa c t_0}, \quad \bar{t} = t - t_0.'$$

(4.4)

In order to derive the transformations relating the measurements of observers in relative motion, he imbedded spacetime into a 5-dimensional pseudo-Euclidean space. Robertson then wrote down the transformations and was satisfied to have found 'the analogue of the Lorentz transformations for the spacetime here considered'.

Restricting himself to cosmologically not very distant objects, but sufficiently far removed so as not to be disturbed by local effects, he found a linear correlation between 'assigned velocity', v, the distance to the nebula, l, and R, which he called the 'radius of the observable world'. He wrote this as

$$v \simeq c \frac{l}{R}. \tag{4.5}$$

Whereas Lemaître had extracted from this formula the velocity–distance relation, later called the 'Hubble law', Robertson derived the size of R from v and l.

In 1929, Robertson returned to the cosmological problem (Robertson 1929). He had in the meantime become aware of Friedmann's work. He split the line element according to space and time: $ds^2 = dt^2 + g_{ij} dx^i dx^j$, where the g_{ij} are functions of time, $g_{ij}(t)$. Then he formalised the approach. He argued that previously the line elements for Einstein's cylindrical and de Sitter's spherical worlds had not been derived from the intrinsic properties of homogeneity and isotropy, and listed a condition 'embodying the uniformity demanded by such a cosmology':

> Assumption I: Space-time shall be spacially homogeneous and isotropic in the sense that it shall admit a transformation which sends an arbitrary configuration in any of the 3-spaces $t = const.$, ... into any other such configuration in the same 3-space in such a way that all intrinsic properties of space-time are left unaltered by the transformation. That is, any such configuration shall be fully equivalent to any other in the same 3-space in the sense that it shall be impossible to distinguish between them by any intrinsic property of space-time.

Further he said:

> Manifolds satisfying this condition alone will be suitable for a cosmology representing the ideal background of the actual universe – provided, of course, that they satisfy Einstein's field equations for some suitable choice of the matter-energy tensor – but they are not of necessity stationary.

Robertson then pointed out that in order to obtain stationary spacetimes, an additional assumption is required:

> Assumption II: There shall exist a transformation of space-time transforming the 3-spaces $t = const.$ among themselves and ... any two of the 3-spaces $t = const.$ shall be fully equivalent in the sense that it shall be impossible to distinguish between them by any intrinsic property of space-time.

Robertson was the first to give a mathematically rigorous formulation of such conditions.

To satisfy Assumption I, he split spacetime into space and time, and wrote the line element for a dynamical universe

$$ds^2 = dt^2 - e^{2f} h_{ij} dx^i dx^j, \tag{4.6}$$

where f is an arbitrary real function of t, and the coefficients h_{ij} are functions of the spatial variables x_1, x_2, x_3. He then expressed his line element in polar coordinates,

$$ds^2 = dt^2 - e^{2f(t)} \left(\frac{dr^2}{1 - r^2/R^2} + r^2 d\theta^2 + r^2 \sin^2 \theta d\varphi^2 \right). \tag{4.7}$$

The universe could also be represented as a general 4-dimensional hypersurface of revolution about z_0, which corresponds to the direction along the time axis

$$z_1^2 + z_2^2 + z_3^2 + z_4^2 = R^2 e^{2f(t)}. \tag{4.8}$$

As in his 1928 publication, he again imbedded the 4-dimensional spacetime in a 5-dimensional pseudo-Euclidean space,

$$ds^2 = dz_0^2 - \left(dz_1^2 + dz_2^2 + dz_3^2 + dz_4^2\right). \tag{4.9}$$

Robertson then turned to the Doppler effect. Because the material content of his idealised universe is at rest, the Doppler shift is calculated from the difference in time of arrival between two flashes of light emitted at times t_0 and $t_0 + \Delta t_0$ from a point source at x_0^i, at rest in the co-moving spatial coordinate system, and arriving at the times t and $t + \Delta t$. For Δt, he obtained from relation (4.7) the expression

$$\Delta t = \Delta t_0 e^{f(t) - f(t_0)}, \tag{4.10}$$

and then derived the expression for the redshift:

$$\Delta \lambda / \lambda = \Delta t / \Delta t_0 - 1. \tag{4.11}$$

He expressed the redshift as a velocity, v, according to the classical Doppler shift, and then related to the time-dependent function $f(t)$ through

$$v = c \times \tanh(f(t) - f(t_0)). \tag{4.12}$$

Robertson commented only briefly on this result. He remarked that of the two stationary universes, only de Sitter's will show such an effect, because in Einstein's universe $f(t)$ is a constant. But in his final deliberation he did not discuss the cosmological significance of $\exp(2f(t))$. In spite of his dynamic formulations he was still hooked on the static universe.

Tolman and the annihilation of matter

In 1929, Richard Tolman joined the theoretical debate about the wavelength shifts in extragalactic nebulae. In his two initial publications he remained firmly within de Sitter's frame, but shifted the emphasis from a 'static' to a 'steady' universe. We have mentioned in the main text that he was puzzled by the contradiction between the tendency of de Sitter's model to scatter the nebulae out of the observable range, and their observed uniform distribution. He therefore explored the possibilities of nebulae disappearing and appearing on de Sitter's horizon (Tolman 1929a,b).

Tolman was not satisfied with his results and, in 1930, he again published several papers on the subject. He had become aware of the dynamic line element of Robertson, and agreed that no static line element could successfully account for the redshift. He then introduced a completely new element into the discussion: the annihilation of matter. He writes the non-static line element as

$$ds^2 = -\frac{e^{2kt}}{\left(1 - r^2/4R^2\right)^2} \left(dx^2 + dy^2 + dz^2\right) + dt^2. \tag{4.13}$$

The constant k in the time-dependent $\exp(2kt)$ is directly related to the rate of transformation of matter into radiation. He justifies this new approach with the argument 'that a general transformation of matter into radiant energy is taking place throughout the universe ... If such a process is going on, the line element for the universe cannot be static, but necessarily must be non-static, since matter and radiation produced from it would not have the same effect on the gravitational field.' (Tolman 1930a, p. 322.) For the relationship between k and the rate of transformation of matter into radiation, he found

$$-\left(\frac{1}{M}\frac{dM}{dt}\right) = 3k. \tag{4.14}$$

Tolman then focuses on the redshift due to the r-dependence of the line element. For a beam of light $ds=0$, and for cosmologically small distances r, $r \ll R$, he can write

$$\frac{dr}{dt} = e^{-kt}. \tag{4.15}$$

This leads to

$$\frac{\lambda + \Delta\lambda}{\lambda} = e^{k(t_2 - t_1)}, \tag{4.16}$$

and, taking $c=1$, and $\Delta r \approx (t_2 - t_1)$ and $\frac{\Delta\lambda}{\lambda} \approx k(t_2 - t_1)$, he obtains the desired redshift

$$\frac{\Delta\lambda}{\lambda} = k \cdot \Delta r. \tag{4.17}$$

The observed linear redshift–distance relationship can be explained as being due to the annihilation of matter.

Through meeting Robertson, and receiving information from de Sitter and Eddington, Tolman had become aware of the work previously done by Friedmann, Lemaître and Robertson. He compared their publications with his own (Tolman 1930c). For that purpose, he re-wrote the non-static line element as

$$ds^2 = -\frac{e^{g(t)}}{(1 - r^2/4R^2)^2}\left(dr^2 + r^2 d\theta^2 + r^2 \sin^2\theta d\phi^2\right) + dt^2. \tag{4.18}$$

R is a constant, and the dependence of the spatial part on time arises from the function $g(t)$. He discussed different treatments of the non-static line element and their impact on redshifts. He concedes to Lemaître that 'the non-static line element can also be applied to an expanding universe without introducing any annihilation of matter'. He sees no conflict between the two mechanisms. Indeed, he finds that annihilation itself may, under appropriate conditions, lead to an expansion of the Universe. To support his claim, he derives a relation between the mass lost in annihilation, the pressure p_0, the density ρ_0, and the metric tensor g:

$$-\frac{1}{M}\frac{dM}{dt} = \frac{6p_0}{\rho_0}\dot{g} + \frac{3}{\rho_0}\frac{dp_0}{dt}. \tag{4.19}$$

The presence of annihilation would either increase the pressure, or increase g, leading to expansion, or it could do both.

5.   Chapter 14: The Einstein–de Sitter universe

In the main text we have described Einstein's conversion to the expanding universe in 1931. Less than one year later, he and de Sitter published what they thought to be the appropriate model, considering the observational situation of the time. They had decided to no longer include the cosmological term in their deliberations, and to opt for a spatially flat universe (Einstein and de Sitter 1932). The line element then simplifies to

$$ds^2 = -R^2(t)(dx^2 + dy^2 + dz^2) + c^2 dt^2,$$
(5.1)

where $R(t)$ is a function of t only. From the field equations, where they neglected the pressure term, they chose

$$\frac{1}{R^2}\left(\frac{dR}{cdt}\right)^2 = \frac{1}{3}\kappa\rho.$$
(5.2)

The term on the left corresponds to the 'Hubble constant', H. With

$$\frac{1}{R}\left(\frac{dR}{cdt}\right) = H,$$
(5.3)

they derived from equation (5.2) the theoretical relation

$$H^2 = \frac{1}{3}\kappa\rho.$$
(5.4)

For $H = 500$ (km/s)/Mpc, the number then generally accepted, they obtained $\rho = 4 \times 10^{-28}$ g/cm^3. Compared with observationally derived densities, this number was rather high. However, they recalled that distances to extragalactic nebulae were still very uncertain, with correspondingly large uncertainties in H and ρ. In that respect they were right. However, they did not receive history's blessing for evicting Λ.

6.   Today's presentation of fundamental cosmological relations

At the beginning of the twenty-first century, cosmological texts give the line element of the 4-dimensional spacetime by the *Friedmann–Lemaître–Robertson–Walker metric*:

$$ds^2 = -dt^2 + R^2(t)\left(\frac{dr^2}{1 - kr^2} + r^2(d\theta^2 + \sin^2\theta d\varphi^2)\right),$$
(6.1a)

where $R(t)$ represents the radius of curvature of the 3-dimensional space. However, R is not restricted to representing a curvature; it may stand for a general scaling factor, which describes the time-dependence of the 3-dimensional space. To show this shift in emphasis, the scaling factor is designated as $a(t)$. Thus

$$ds^2 = -dt^2 + a^2(t)\left(\frac{dr^2}{1 - kr^2} + r^2(d\theta^2 + \sin^2\theta d\varphi^2)\right).$$
(6.1b)

The scaling factor $a(t)$ can be determined from Einstein's fundamental equations. Three different values of k distinguish three different geometries of space:

$k < 0$: negative curvature of space; it corresponds to the curved, open universe.

$k = 0$: no curvature of space; it corresponds to a curved spacetime where the 3-dimensional space is flat.

$k > 0$: positive curvature of space; it corresponds to the curved, closed universe.

Without loss of generality, we can set either $k=-1$, $k=0$, or $k=+1$, and we define the spatial part as

$$d\sigma^2 = \frac{dr^2}{1-kr^2} + r^2 \left(d\theta^2 + \sin^2\theta d\varphi^2\right). \tag{6.2}$$

The line element (6.1b) is then written as

$$ds^2 = -dt^2 + a^2(t)d\sigma^2. \tag{6.3}$$

The case $k=0$:

$$d\sigma^2 = dr^2 + r^2\left(d\theta^2 + \sin^2\theta d\varphi^2\right) = dx^2 + dy^2 + dz^2. \tag{6.4}$$

This is the flat (Euclidean) space.

The case $k=+1$: we substitute $r=\sin\chi$, and obtain

$$d\sigma^2 = d\chi^2 + \sin^2\chi\left(d\theta^2 + \sin^2\theta d\varphi^2\right). \tag{6.5}$$

This is the metric of a 3-dimensional sphere, thus spacetime corresponds to a closed universe.

The case $k=-1$: we substitute $r=\sinh\chi$, and obtain

$$d\sigma^2 = d\chi^2 + \sinh^2\chi\left(d\theta^2 + \sin^2\theta d\varphi^2\right). \tag{6.6}$$

This is the metric of a 3-dimensional space of constant negative curvature. It corresponds to the curved open universe.

The Universe can be modelled as a perfect fluid, characterised by its rest frame energy density ρ and isotropic pressure p. Thus it has no viscosity or shear stresses. Examples are universes filled with ordinary stars, or galaxies; they are practically collisionless, thus pressure is negligible compared with the energy density. Another important example of a perfect fluid is a universe filled with radiation. This includes electromagnetic radiation on the one hand, and massive particles with relative velocities close to the speed of light on the other. These two types of perfect fluids – dust and radiation – represent the most relevant models in cosmology.

The energy–momentum tensor of a perfect fluid is

$$T_{ij} = (p+\rho)u_i u_j + pg_{ij}. \tag{6.7}$$

p and ρ are measured in the rest frame, u_i is the four-velocity of the fluid. In co-moving coordinates the fluid is at rest, therefore the velocity is given by $u_i=(0,0,0,1)$. The energy–momentum tensor reads

$$T^i_j = \mathrm{diag}(p,p,p,-\rho). \tag{6.8}$$

Taking the trace yields

$$\mathrm{tr}T = T = T^i_i = -\rho + 3p. \tag{6.9}$$

From the earlier introduced condition, $\nabla_j T^{ij} = 0$, we derive

$$\frac{\dot{\rho}}{\rho} = -3(1+w)\frac{\dot{a}}{a}, \tag{6.10}$$

where w is a constant, independent of time, coming from the equation of state for a perfect fluid,

$$p = w\rho. \tag{6.11}$$

The equation for $\dot{\rho}/\rho$ is then integrated and yields

$$\rho \sim a^{-3(1+w)}. \tag{6.12}$$

Let us focus on the two types of a perfect fluid: dust and radiation. We distinguish between matter-dominated and radiation-dominated universes. In a matter-dominated universe, the energy density is mostly due to dust, whereas in a radiation-dominated universe, most of the energy density is due to radiation.

Matter-dominated universe

Here, particles dominate over radiation, thus, pressure disappears ($w=0$), and we obtain

$$\rho \sim a^{-3}. \tag{6.13}$$

With the expansion of the Universe, the energy density ρ decreases with the third power of the expansion parameter.

Radiation-dominated universe

In a radiation dominated fluid, the equation of state gives $3p = \rho$, thus $w = 1/3$, and therefore

$$\rho \sim a^{-4}. \tag{6.14}$$

Here the energy density decreases with the fourth power of the expansion parameter. This is obvious from the fact that the number density of photons decreases just as the particle density but, in addition, each photon loses energy due to the wavelength shift towards the red. With the two different behaviours there must be a crossover in the expanding universe, from the early radiation-dominated to today's matter-dominated state.

We now turn towards the vacuum-dominated universe. For that purpose we write Einstein's equation (1.1b), including the cosmological term Λ, as

$$G_{ij} = 8\pi G T_{ij} - \Lambda g_{ij}. \tag{6.15}$$

We now let Λ become part of the energy–momentum tensor of the vacuum,

$$G_{ij} = 8\pi G \left(T_{ij} + T_{ij}^{(vac)} \right), \tag{6.16}$$

with

$$T_{ij}^{(vac)} = -\frac{\Lambda}{8\pi G} g_{ij}. \tag{6.17}$$

This energy–momentum tensor has the same form as for the perfect fluid, with energy density $\rho^{(vac)}$ and pressure $p^{(vac)}$, and $w = -1$:

$$\rho^{(vac)} = -p^{(vac)} = \frac{\Lambda}{8\pi G}. \tag{6.18}$$

The cosmological term Λ results in a positive vacuum energy density $\rho^{(vac)}$. Lemaître (1934) had already pointed out this fact. The vacuum energy density has a repulsive effect, as intended by Einstein. The effect does not depend on the expansion coefficient, a. This is a crucial point for the long-term history of the Universe. As both matter and radiation energy densities decrease with an increase in a, the vacuum energy expressed in Λ will dominate the future expansion. However, whether Λ is indeed a constant is still debated.

We recall that Einstein's fundamental equations cannot be solved in full generality. The Friedmann equations have been derived from Einstein's equations assuming homogeneity and isotropy for a fluid with density ρ and pressure p. They are now written in the form

$$\frac{\ddot{a}}{a} = -\frac{4\pi G}{3}\left(\rho + \frac{3p}{c^2}\right) + \frac{\Lambda}{3} \tag{6.19}$$

$$\left(\frac{\dot{a}}{a}\right)^2 = \frac{8\pi G}{3}\rho - \frac{kc^2}{a^2} + \frac{\Lambda}{3}. \tag{6.20}$$

These equations form an essential part of modern cosmology. Metrics of the form (6.1b), which obey the Friedmann equations, and are therefore also solutions of Einstein's equations, are called Friedmann–Lemaître–Robertson–Walker (FLRW) universes. Thus, Friedmann's equations give the constraints on the metrics in Eq. (6.1b). In addition, we define the expansion parameter H, the deceleration parameter q, the critical density ρ_{crit}, and the density parameter Ω:

$$H = \frac{\dot{a}}{a}, \quad q = -a\frac{\ddot{a}}{\dot{a}^2}, \quad \rho_{crit} = \frac{3H^2}{8\pi G}, \quad \Omega = \frac{\rho}{\rho_{crit}} = \frac{8\pi G}{3H^2}. \tag{6.21}$$

The Friedmann equation (6.20) can then be written as

$$\Omega - 1 = \frac{k}{1 + 2a^2}. \tag{6.22}$$

This relation determines the sign of k, and it tells us whether the Universe is closed, flat or open:

$\rho < \rho_{crit} \rightarrow \Omega < 1 \rightarrow k = -1$: open universe,
$\rho = \rho_{crit} \rightarrow \Omega = 1 \rightarrow k = 0$: open universe,
$\rho > \rho_{crit} \rightarrow \Omega > 1 \rightarrow k = +1$: closed universe.

As the density parameter can in principle be obtained from observations, observations can tell us whether we live in an open, a flat or a closed universe.

The future of our Universe looks different for different values of k. For the open and the flat universe, $k < 0$ and $k = 0$, relation (6.20) shows that the expansion velocity will always remain positive, unless $\Lambda = 0$ and ρ approaches zero. In this latter case, the Universe still lives an infinitely long time, but the expansion tends asymptotically to a halt, corresponding to the Einstein–de Sitter universe.

For $k = +1$, we have to distinguish between $\Lambda = 0$ and $\Lambda > 0$. From observations, we know that at present $\dot{a} > 0$. In addition, we can neglect pressure, thus $p = 0$. We now look at relation (6.19). For $\Lambda = 0$, the expansion is decelerating, and will eventually turn into a contraction. For $\Lambda > 0$, the Universe will continue to expand. If the expansion should be decelerating at the present time, which according to observations seems not to be the case, this will turn into a future acceleration with the diminution of ρ.

Abbreviations

ApJ	The Astrophysical Journal
BAN	Bulletin of the Astronomical Institutes of the Netherlands
MNRAS	Monthly Notices of the Royal Astronomical Society
PASP	Publications of the Astronomical Society of the Pacific
Phil. Trans.	Philosophical Transactions of the Royal Society of London
PNAS	Proceedings of the National Academy of Sciences of the United States of America
SAW	Sitzungsberichte der Königlich Preußischen Akademie der Wissenschaften (Berlin)

References

Baade, W. (1956). The period-luminosity relation of the Cepheids. *PASP*, **68**, 5–16.

Bentley, R. (1693). *A Confutation of Atheism*. Annenberg Rare Book and Manuscript Library. BL2747.3. B46 1693. Schoenberg Center for Electronic Text & Image.

Berendzen, R. and Hoskin, M. (1971). Hubble's announcement of Cepheids in spiral nebulae. *Astronomical Society of the Pacific Leaflets*, **10**, 425–440.

Bohlin, K. (1909). On the galactic system with regard to its structure, origin, and relations in space. *Kungl. Svenska Vetenskapsakademiens handlingar*, **43**, No. 10, 2–29.

Bondi, H. and Gold, T. (1948). The steady-state theory of the expanding universe. *MNRAS*, **108**, 252–270.

Bruno, G. (1584). *The Ash Wednesday Supper*. Edited and translated by E. A. Gosseli and L. S. Lerner. North Haven, CT: Archon Books, 1977.

Brush, S. G. (1996). *Transmuted Past*. Cambridge: Cambridge University Press.

Burbidge, E. M., Burbidge, G. R., Fowler, W. A. and Hoyle, F. (1957). Synthesis of the elements in stars. *Reviews of Modern Physics*, **29**, 547–650.

Charlier, C. V. I. (1922). How an infinite world may be built up. *Arkiv för matematik, astronomi och fysik*, **16**, No.22, 1–34.

Chéseaux, J. P. Loys de (1744). *Traité de la Comete*. Lausanne & Genève: Marc-Michel Bousquet et Co.

Clerke, Agnes M. (1890). *The System of the Stars*. London: Longmans, Green, and Co.

Colless, M. *et al.* (2001). The 2dF Galaxy Redshift Survey: spectra and redshifts. *MNRAS*, **328**, 1039–1063.

Condon, E. (1925). The age of the stars. *PNAS*, **11**, 125–130.

Copernicus, N. (1543). *De Revolutionibus Orbium Coelestium. Nürnberg*. Translated by E. Rosen. London: The Warnock Library, 1978.

CPAE (1998). *The Collected Papers of Albert Einstein*. **8B**, Princeton, NJ: Princeton University Press.

CPAE (2002). *The Collected Papers of Albert Einstein*. **7**, Princeton, NJ: Princeton University Press.

Curtis, H. D. (1917). Novae in spiral nebulae and the island universe theory. *PASP*, **29**, 206–207.

Cusa, N. de (1440). *De docta ignorantia*. Latin/German. Buch I und II. Hamburg: Felix Meiner, 1970/1967.

Derham, W. (1733). Observations of the appearances among the fix'd stars, called nebulous stars. *Phil. Trans.*, **38**, 70–74.

Descartes, R. (1668). *Les principes de la philosophie*. Paris: Bobin & Le Gras (MPIWG Digital Rare Book Library).

de Sitter, W. (1917). On Einstein's theory of gravitation and its astronomical consequences. *MNRAS*, **78**, 3–28.

de Sitter, W. (1930a). Proceeding of the R.A.S. *The Observatory*, **53**, 37–39.

de Sitter, W. (1930b). On the magnitudes, diameters and distances of the extragalactic nebulae, and their apparent radial velocities. *BAN*, **185**, 157–171.

de Sitter, W. (1930c). The expanding universe. Discussion of Lemaître's solution of the equations of the inertial field. *BAN*, **193**, 211–218.

de Sitter, W. (1931). Contribution to a British Association discussion on the evolution of the Universe. *Nature*, **128**, 706–709.

Dicke, R. H., Peebles, P. J. E., Roll, P. G. and Wilkinson, D. T. (1965). Cosmic black-body radiation. *ApJ*, **142**, 414–419.

Digges, L., (1576). *A prognostication everlastinge corrected and augmented by Thomas Digges*. The English Experience. Facsimile. Norwood, NJ: Walter J. Johnson, 1975.

Dingle, H. (1933a). The age of the Universe and its bearing on astronomical problems. *PASP*, **45**, 159–170.

Dingle, H. (1933b). On E. A. Milne's theory of world structure and the expansion of the Universe. *Zeitschrift für Astrophysik*, **7**, 167–179.

Dingle, H. (1953). Science and modern cosmology. *MNRAS*, **113**, 393–407.

Duerbeck, H. W. and Seitter, W. (2000). In Edwin Hubbles Schatten: Frühe Arbeiten zur Expansion des Universums. *Beiträge zur Astronomiegeschichte*, **3**, 120–147.

Du Val, P. (1924). Geometrical note on de Sitter's world. *London, Edinburgh and Dublin Philosophical Magazine* (Ser. 6), **47**, 930–938.

Dyson, F. W., Eddington, A. S. and Davidson, C. (1920). A determination of the deflection of light by the Sun's gravitational field, from observations made at the total eclipse of May 20, 1919. *Phil. Trans.*, **220**, 291–333.

Eddington, A. S. (1920). The internal constitution of the stars. *The Observatory*, **43**, 341–358.

Eddington, A. S. (1921). A generalisation of Weyl's theory of the electromagnetic and gravitational fields. *Proceedings of the Royal Society of London*, **99**, 104–122.

Eddington, A. S. (1923/1924). *The Mathematical Theory of Relativity*. Cambridge: Cambridge University Press, 1st and 2nd Edn.

Eddington, A. S. (1924). Radial velocities and the curvature of space-time. *Nature*, **113**, 746–747.

Eddington, A. S. (1925). *Relativitätstheorie in mathematischer Behandlung*. Berlin: Julius Springer.

Eddington, A. S. (1930). On the instability of Einstein's spherical world. *MNRAS*, **90**, 668–678.

Eddington, A. S. (1931). The end of the world: from the standpoint of mathematical physics. *Nature*, **127**, 447.

Eddington, A. S. (1933). *The Expanding Universe*. Cambridge: Cambridge University Press.

Eddington, A. S. (1940). Forty years of astronomy. In *Background to Modern Science* Cambridge: The University Press.

Einstein, A. (1915a). Zur Allgemeinen Relativitätstheorie. *SAW*, 778–786.

Einstein, A. (1915b). Die Feldgleichungen der Gravitation. *SAW*, 844–847.

Einstein, A. (1916). Die Grundlage der allgemeinen Relativitätstheorie. *Annalen der Physik*, **49**, 769–822.

Einstein, A. (1917). Kosmologische Betrachtungen zur allgemeinen Relativitätstheorie, *SAW*, 142–152.

Einstein, A. (1918a). Kritisches zu einer von Hrn. De Sitter gegebenen Lösung der Gravitationsgleichungen. *SAW*, 270–272.

Einstein, A. (1918b). Der Energiesatz in der allgemeinen Relativitätstheorie. *SAW*, 448–459.

Einstein, A. (1918c). Prinzipielles zur allgemeinen Relativitätstheorie. *Annalen der Physik*, **55**, 241–244.

Einstein, A. (1922). Bemerkung zu der Arbeit von A. Friedman 'Über die Krümmung des Raumes'. *Zeitschrift für Physik*, **11**, 326.

Einstein, A. (1923a). Notiz zu der Arbeit von A. Friedman 'Über die Krümmung des Raumes'. *Zeitschrift für Physik*, **16**, 228.

Einstein, A. (1923b). *Notiz zu der Arbeit von A. Friedmann*. The Albert Einstein Archives, The Hebrew University of Jerusalem. Archivnummer 1–26.00.

Einstein, A. (1930). *Tagebuch Amerika-Reise 1930*. The Albert Einstein Archives, The Hebrew University of Jerusalem. Archivnummer 29–134.

Einstein, A. (1931a). *Letter to Besso*. The Albert Einstein Archives, The Hebrew University of Jerusalem. Archivnummer 7–125.

Einstein, A. (1931b). Kosmologische Betrachtungen zur allgemeinen Relativitätstheorie. *SAW*, 235–237.

Einstein, A. (1931c). *Letter to Tolman*. The Albert Einstein Archives, The Hebrew University of Jerusalem. Archivnummer 23–30.

Einstein, A. (1932). *Tagebuch Amerika-Reise 1931/32*. The Albert Einstein Archives, The Hebrew University of Jerusalem. Archivnummer 29–136.

Einstein, A. and de Sitter, W. (1932). On the relation between the expansion and the mean density of the Universe. *PNAS*, **18**, 213–214.

Fraunhofer, J. (1814/15). Bestimmung des Brechungs- und Farbenzerstreuungs-Vermögens verschiedener Glasarten, in Bezug auf die Vervollkommnung achromatischer Fernröhren. *Denkschriften der Königlichen Academie der Wissenschaften zu München*, **5**, 193–225.

Friedmann, A. (1922). Über die Krümmung des Raumes. *Zeitschrift für Physik*, **10**, 377–386.

Friedmann, A. (1923). *Die Welt als Raum und Zeit*. Ostwalds Klassiker der exakten Wissenschaften, Harri Deutsch, 2006.

Friedmann, A. (1924). Über die Möglichkeit einer Welt mit konstanter negativer Krümmung des Raumes. *Zeitschrift für Physik*, **21**, 326–332.

Galilei, G. (1610). *Sidereus Nuncius*. Venice.

Gamow, G. (1948). The evolution of the Universe. *Nature*, **162**, 680–682.

Geller, M. J. and Huchra, J. P. (1989). Mapping the Universe. *Science*, **246**, 897–903.

Gingerich, O. (2002). *An annotated census of Copernicus' De Revolutionibus (Nuremberg, 1543 and Basel, 1556)*. Leiden: Brill.

Greenstein, J. L. (1938). The temperatures of the extragalactic nebulae and the redshift correction. *ApJ*, **88**, 605–617.

Gregory, S. A. and Thompson, L. A. (1978). The Coma/A1367 supercluster and its environs. *ApJ*, **222**, 784–799.

Gregory, S. A. and Thompson, L. A. (1982). Superclusters and voids in the distribution of galaxies. *Scientific American*, **246**, 88–96.

Haas, A. (1925). *Introduction to Theoretical Physics, II*. London: Constable & Company Ltd.

Halley, E. (1716). An account of several nebulae or lucid spots like clouds, lately discovered among the fixt stars by help of the telescope. *Phil. Trans.*, **29**, 390–392.

Heckmann, O. (1932). Die Ausdehnung der Welt in ihrer Abhängigkeit von der Zeit. *Veröffentlichungen der Universitäts-Sternwarte zu Göttingen*, **23**, 181–190.

Herschel, W. (1784). Account of some observations tending to investigate the construction of the heavens. *Phil. Trans.*, **74**, 437–451.

Herschel, W. (1785). On the construction of the heavens. *Phil. Trans.*, **75**, 213–266.

Herschel, W. (1786). Catalogue of one thousand new nebulae and clusters of stars. *Phil. Trans.*, **76**, 457–499.

Herschel, W. (1791). On nebulous stars, properly so called. *Phil. Trans.*, **81**, 71–88.

Herschel, W. (1811). Astronomical observations relating to the construction of the heavens, arranged for the purpose of a critical examination, the result of which appears to throw some new light upon the organization of the celestial bodies. *Phil. Trans.*, **101**, 269–336.

Herschel, W. (1814). Astronomical observations relating to the sidereal part of the heavens, and its connection with the nebulous part; arranged for the purpose of a critical examination. *Phil. Trans.*, **104**, 248–284.

Hevelius, J. (1687). *Catalogus stellarum fixarum*. Gedani: typis Johannis Zachariae Stollii.

Hevelius, J. (1690). *Firmamentum Sobiescianum, Uranographia*. Gedani: typis Johannis Zachariae Stollii.

Hooke, R. (1674). An attempt to prove the motion of the Earth by observations. In 'The Cutler lectures of Robert Hooke'; reprinted by R. T. Gunter: *Early Science in Oxford*. Oxford 1931.

Hoskin, M. (1976). The 'Great Debate': what really happened. *Journal for the History of Astronomy*, **7**, 169–182.

Hoyle, F. (1948). A new model for the expanding universe. *MNRAS*, **108**, 372–382.

Hoyle, F. and Sandage, A. (1956). The second-order term in the redshift–magnitude relation. *PASP*, **68**, 301–307.

Hoyle, F. and Tayler, R. J. (1964). The mystery of the cosmic helium abundance. *Nature*, **203**, 1108–1110.

Hubble, E. (1925a). NGC 6822, a remote stellar system. *ApJ*, **62**, 409–433.

Hubble, E. (1925b). Cepheids in spiral nebulae. *Popular Astronomy*, **33**, 158–159, 252–255.

Hubble, E. (1925c). Cepheids in spiral nebulae. *The Observatory*, **48**, 139–142.

Hubble, E. (1926). Extra-galactic nebulae. *ApJ*, **64**, 321–369.

Hubble, E. (1929a). A relation between distance and radial velocity among extra-galactic nebulae. *PNAS*, **15**, 168–173.

Hubble, E. (1929b). A clue to the structure of the Universe. *Astronomical Society of the Pacific Leaflets*, **23**, 93–96.

Hubble, E. (1930). *Letter to W. de Sitter*. August 21, 1930. Henry E. Huntington Library archives, HUB 616.

Hubble, E. (1931). *Letter to W. de Sitter*. Henry E. Huntington Library archives, HUB 617.

Hubble, E. (1935). Angular rotations of spiral nebulae. *ApJ*, **81**, 334–335.

Hubble, E. (1936). *The Realm of the Nebulae*. Yale: Yale University Press.

Hubble, E. (1937). *The Observational Approach to Cosmology*. Oxford: The Clarendon Press.

Hubble, E. (1953). The law of red-shifts. *MNRAS*, **113**, 658–666.

Hubble, E. and Humason, M. L. (1931). The velocity–distance relation among extra-galactic nebulae. *ApJ*, **74**, 43–80.

Huggins, W. (1864). On the spectra of some of the nebulae. *Phil. Trans.*, **154**, 437–444.

Huggins, W. (1866a). Further observations on the spectra of some of the nebulae, with a mode of determining the brightness of these bodies. *Phil. Trans.*, **156**, 381–397.

Huggins, W. (1866b). *Analyse spectrale des corps célestes*. Paris: Gautier-Villars.

Humason, M. L. (1929). The large radial velocity of NGC 7619. *PNAS*, **15**, 167.

Humason, M. L. (1931). Apparent velocity shifts in the spectra of faint nebulae. *ApJ*, **74**, 35–42.

Humason, M. L. (1936). The apparent velocities of 100 extra-galactic nebulae. *ApJ*, **83**, 10–22.

Humason, M. L., Mayall, N. U. and Sandage, A. R. (1956). Redshifts and magnitudes of extragalactic nebulae. *Astronomical Journal*, **61**, 97–162.

Jeans, J. H. (1919). *Problems of Cosmogony and Stellar Dynamics*. Cambridge: Cambridge University Press.

Jeans, J. H. (1928). *Astronomy and Cosmogony*. Cambridge: Cambridge University Press.

Jeans, J. H. (1931). Contribution to a British Association discussion on the evolution of the Universe. *Nature*, **128**, Supplement, 701–704.

Kant, I. (1755). *Allgemeine Naturgeschichte und Theorie des Himmels oder Versuch von der Verfassung und dem mechanischen Ursprunge des ganzen Weltgebäudes, nach Newtonischen Grundsätzen abgehandelt*.

Kapteyn, J. C. and van Rhijn, P. J. (1920). On the distribution of the stars in space especially in the high galactic latitudes. *ApJ*, **52**, 23–38.

Kepler, J. (1596). *Mysterium Cosmographicum*. Tübingen.

Kepler, J. (1610). *Dissertatio cum nuncio sideris Galilei. Faksimilewiedergabe*. Münchern: Poerschke & Weiner, 1964.

Kepler, J. (1618–21). *Epitome of Copernican Astronomy*. Translated by Ch. G. Wallis. Amherst, NY: Prometheus Books, 1995.

Kopff, A. (1921). Die Untersuchungen H. Shapleys über Sternhaufen und Milchstraßensystem. *Die Naturwissenschaften*, **9**, 769–774.

Kragh, H. and Smith R. W. (2003). Who discovered the expanding universe? *History of Science*, **41**, 141–162.

Lambert, D. (2000). *Un atome d'univers: La vie et l'oeuvre de Georges Lemaître*. Bruxelles: Éditions Lessius.

Lanczos, K. (1922). Bemerkungen zur de Sitterschen Welt. *Physikalische Zeitschrift*, **23**, 539–543.

Lanczos, K. (1924). Über eine stationäre Kosmologie im Sinne der Einsteinschen Gravitationstheorie. *Zeitschrift für Physik*, **21**, 73–110.

Laplace, P. S. (1796). *Exposition du système du monde*. Paris: Gauthier-Villars, 1884. Document électronique, Institut National de la Langue Francaise.

Lemaître, G. (1925a). Note on de Sitter's universe. *Journal of Mathematics and Physics*, **4**, 188.

Lemaître, G. (1925b). Note on de Sitter's universe. *Physical Review*, **25**, Ser.II, 903.

Lemaître, G. (1925c). *Letter to Eddington* (November 29, 1925). Archives Lemaître, Université Catholique de Louvain.

Lemaître, G. (1927). Un univers homogène de masse constante et de rayon croissant, rendant compte de la vitesse radiale des nébuleuses extra-galactiques. *Annales de la Société scientifique de Bruxelles*, Série A, **47**, 49–59.

Lemaître, G. (1929). La grandeur de l'espace. *Revue des Questions scientifiques*, **15**, 189–216.

Lemaître, G. (1930a). *Letter to Eddington* (early 1930). Archives Lemaître, Université Catholique de Louvain.

Lemaître, G. (1930b). On the random motion of particles in the expanding universe. Explanation of a paradox. *BAN*, **200**, 273–274.

Lemaître, G. (1931a). A homogeneous universe of constant mass and increasing radius accounting for the radial velocity of extra-galactic nebulae. *MNRAS*, **91**, 483–490.

Lemaître, G. (1931b). The expanding universe. *MNRAS*, **91**, 490–501.

Lemaître, G. (1931c). The beginning of the world from the point of view of quantum theory. *Nature*, **127**, 706.

Lemaître, G. (1931d). L'expansion de l'espace. *Revue des Questions scientifiques*, **20**, 391–410.

Lemaître, G. (1931e). Contribution to a British Association discussion on the evolution of the Universe. *Nature*, **128**, Supplement, 704–706.

Lemaître, G. (1933). L'univers en expansion. *Annales de la Société scientifique de Bruxelles*. Série A, **53**, 51–85.

Lemaître, G. (1934). Evolution of the expanding universe. *PNAS*, **20**, 12–17.

Lemaître, G. (1947). *L'hypothèse de l'atome primitif*. Les Conférences du Palais de la Découverte. Paris: Université de Paris.

Lemaître, G. (1950). L'expansion de l'Univers, par Paul Couderc. *Annales d'Astrophysique*, **13**, 344–345.

Lemaître, G. (1958). Rencontres avec A. Einstein. *Revue des Questions Scientifiques*, **129**, 129–132.

Lemaître, G. (1967). L'expansion de l'Univers. *Revue des Questions Scientifiques*, **138**, 153–162.

Lubin, L. M. and Sandage, A. (2001). The Tolman surface brightness test for the reality of the expansion. IV. A measurement of the Tolman signal and the luminosity evolution of early-type galaxies. *Astronomical Journal*, **122**, 1084–1103.

Lundmark, K. (1924). The determination of the curvature of space-time in de sitter's world. *MNRAS*, **84**, 747–764.

Lundmark, K. (1925). The motions and distances of spiral nebulae. *MNRAS*, **85**, 865–894.

Maeyama, Y. (2002). Tycho Brahe's stellar observations. An accuracy test. Tycho Brahe and Prague: Crossroads of European Science. *Acta Historica Astronomiae*, **16**, 113–119.

Marius, S. (1614). *Mundus Iovialis*. Latin/German version, ed. Joachim Schlör. Gunzenhausen: Schrenk-Verlag, 1988.

Mattig, W. (1958). Über den Zusammenhang zwischen Rotverschiebung und scheinbarer Helligkeit. *Astron. Nachrichten*, **284**, 109–111.

Mattig, W. (1959). Über den Zusammenhang zwischen der Anzahl der extragalaktischen Objekte und der scheinbaren Helligkeit. *Astron. Nachrichten*, **285**, 1–2.

McCrea, W. H. and McVittie, G. C. (1930). On the contraction of the Universe. *MNRAS*, **91**, 128–133.

McCrea, W. H. and McVittie, G. C. (1931). The expanding universe. *MNRAS*, **92**, 7–12.

McVittie, G. C. (1931). The problem on n bodies and the expansion of the Universe. *MNRAS*, **91**, 274–283.

McVittie, G. C. (1967). Georges Lemaître. *Quarterly Journal of the Royal Astronomical Society*, **8**, 294–297.

Messier, C. (1774). Catalogue des Nébuleuses & des amas d'Étoiles, que l'on découvre parmi les Étoiles fixes sur l'horizon de Paris; observées à l'Observatoire de la Marine, avec differens instruments. *Mémoires de l'Académie Royale des Sciences for 1771*, Paris (dated February 16, 1771, published 1774), p. 435–461 + Pl. VIII [Bibcode: 1774MmARS1771..435M].

Milne, E. A. (1933). World-structure and the expansion of the Universe. *Zeitschrift für Astrophysik*, **6**, 1–95.

Milne, E. A. (1935). *Relativity, Gravitation, and World-Structure*. Oxford: The Clarendon Press.

Murphy, N. and Ellis, G. F. R. (1996). *On the Moral Nature of the Universe: Theology, Cosmology, and Ethics*. Minneapolis, MI: Fortress Press.

Newton, I. (1692). *Letter One to Richard Bentley: Philosophical Writings*. Edited by A. Janiak. Cambridge: Cambridge University Press, 2004.

Newton, I. (1726). *The Principia, Mathematical Principles of Natural Philosophy*. Translated by I. B. Cohen and A. Whitman. Berkeley, CA: University of California Press, 1999.

Newton, I. (1730). *Opticks*. New York: Dover Publications, 1952.

Oke, J. B. and Sandage, A. (1968). Energy distributions, K corrections, and the Stebbins–Whitford effect for giant elliptical galaxies. *ApJ*, **154**, 21–32.

Olbers, H. (1823). Über die Durchsichtigkeit des Weltraumes. *Astronomisches Jahrbuch für das Jahr 1826*. Ed. J. E. Bode, Berlin.

Öpik, E. (1922). An estimate of the distance of the Andromeda nebula. *ApJ*, **55**, 406–410.

Oleak, H. (1995). Scheiners Spektrum des Andromedanebels. Über die Natur der Spiralnebel. *Die Sterne*, **71**, 95–100.

Peebles, P. J. E. (1966). Primordial helium abundance and the primordial fireball. II. *ApJ*, **146**, 542–552.

Penzias, A. A. and Wilson, R. W. (1965). A measurement of excess antenna temperature at 4080 Mc/s. *ApJ*, **142**, 419–421.

Perlmutter, S. *et al.* (1998). Discovery of a supernova explosion at half the age of the Universe. *Nature*, **391**, 51–54.

Perrin, J. (1919). Matière et Lumière. *Annales de Physique*, **11**, B Ser. 9. 5–108.

Puiseux, P. (1912). Les Nébuleuses Spirales. *Revue Scientifique*, **14**, 417–422.

Riess, A. G. *et al.* (1998). Observational evidence from supernovae for an accelerating Universe and a cosmological constant. *Astronomical Journal*, **116**, 1009–138.

Robertson, H. P. (1928). On relativistic cosmology. *Phil. Mag.*, Ser. 7, Vol. **5**, Supplement, May 1928, 835–848.

Robertson, H. P. (1929). On the foundations of relativistic cosmology. *PNAS*, **15**, 822–829.

Robertson, H. P. (1933). Relativistic cosmology. *Reviews of Modern Physics*, **5**, 62–90.

Robertson, H. P. (1938). The apparent luminosity of a receding nebula. *Zeitschrift für Astrophysik*, **15**, 69–81.

Robertson, H. P. (1955). The theoretical aspects of the nebular redshift. *PASP*, **67**, 82–98.

Rosse, Earl of (1850). Observations on the nebulae. *Phil. Trans.*, **140**, 499–514.

Rosse, Earl of (1861). On the Construction of specula of six-feet aperture; and a selection from the observations of nebulae made with them. *Phil. Trans.*, **151**, 681–745.

Sandage, A. (1958). Current problems in the extragalactic distance scale. *ApJ*, **127**, 513–526.

Sandage, A. (1961). The ability of the 200-inch telescope to discriminate between selected world models. *ApJ*, **133**, 355–392.

Sandage, A. R. (1970). Cosmology: a search for two numbers. *Physics Today*, **23**, 34–41.

Sandage, A. (2000). Episodes in the discovery of variations in the chemical composition of stars and galaxies. *PASP*, **112**, 293–296.

Sandage, A. (2004). Observational cosmology II: The expansion of the Universe and the curvature of space. In *Centennial History of the Carnegie Institution of Washington*, Vol. I. Cambridge: Cambridge University Press.

Sandage, A. and Perelmuter, J. M. (1991). The surface brightness test for the expansion of the Universe. III – Reduction of data for the several brightest galaxies in clusters to standard conditions and a first indication that the expansion is real. *ApJ*, **370**, 455–473.

Scheiner, J. (1899). On the spectrum of the great nebula in Andromeda. *ApJ*, **9**, 149–150.

Schrödinger, E. (1918). Über ein Lösungssystem der allgemein kovarianten Gravitationsgleichungen. *Physikalische Zeitschrift*, **19**, 20–22.

Schwarzschild, K. (1900). Ueber das zulässige Krümmungsmass des Raumes. *Vierteljahrschrift der Astronomischen Gesellschaft*, **35**, 337–347.

Seitter, W. C. and Duemmler, R. (1989). The cosmological constant – historical annotations. In *Morphological Cosmology*, edited by P. Flin and H. Duerbeck. Berlin: Springer, p. 377–387.

Shapley, H. (1916). Studies based on the colors and magnitudes in stellar clusters. II. Thirteen hundred stars in the Hercules Cluster (Messier 13). *Contributions from the Mount Wilson Solar Observatory*, **116**, 3–92.

Shapley, H. (1917). Color-indices of stars in the galactic clouds. *Contributions from the Mount Wilson Solar Observatory*, **133**, 1–12. Reprinted in *ApJ*. 46, 64–75 (1917).

Shapley, H. (1918a). A summary of results bearing on the structure of the sidereal universe. *PNAS*, **4**, 224–229.

Shapley, H. (1918b). Globular clusters and the structure of the galactic system. *PASP*, **30**, 42–54.

Shapley, H. (1918c). Remarks on the arrangement of the sidereal universe. *Contributions from the Mount Wilson Solar Observatory*, **157**, 1–26. Reprinted in *ApJ*. 49, 311–336 (1919).

Shapley, H. (1919). On the existence of external galaxies. *PASP*, **31**, 261–267.

Shapley, H. (1969). *Through Rugged Ways to the Stars*. New York: Charles Scribner's Sons.

Shapley, H. and Shapley, M. B. (1919). Studies based on the colors and magnitudes in stellar clusters. *ApJ*, **50**, 107–140.

Silberstein, L. (1924). The curvature of de Sitter's space-time derived from globular clusters. *MNRAS*, **84**, 363–366.

Slipher, V. M. (1913). The radial velocity of the Andromeda nebula. *Lowell Observatory Bulletin*, No. 58.

Slipher, V. M. (1915). Spectrographic observations of nebulae. *Popular Astronomy*, **23**, 21–24.

Slipher, V. M. (1917). Nebulae. *Proc. American Philosophical Soc.*, **56**, 403–409.

Stenflo, J. O. (1970). Hale's attempts to determine the Sun's general magnetic field. *Solar Physics*, **14**, 263–273.

Straumann, N. (2007). Dark energy. In *Approaches to Fundamental Physics: An Assessment of Current Theoretical Ideas*, edited by E. Seiler and I. O. Stamatescu, Lecture Notes in Physics. Berlin: Springer, p. 327–398.

Strohmaier, G. (1984). *Die Sterne des Abd ar-Rahman as-Sufi*. Leipzig und Weimar: Gustav Kiepenheuer Verlag.

Strömberg, G. (1925). Analysis of radial velocities of globular clusters and non-galactic nebulae. *ApJ*, **61**, 353–388.

Strömgren, B. (1933). On the interpretation of the Hertzsprung–Russell diagram. *Zeitschrift für Astrophysik*, **7**, 222–248.

Tolman, R. C. (1929a). On the possible line elements for the Universe. *PNAS*, **15**, 297–304.

Tolman, R. C. (1929b). On the astronomical implications of the de Sitter line element for the Universe. *ApJ*, **69**, 245–274.

Tolman, R. C. (1930a). The effect of the annihilation of matter on the wave-length of light from the nebulae. *PNAS*, **16**, 320–337.

Tolman, R. C. (1930b). More complete discussion of the time-dependence of the non-static line element for the Universe. *PNAS*, **16**, 409–420.

Tolman, R. C. (1930c). Discussion of various treatments which have been given to the non-static line element for the Universe. *PNAS*, **16**, 582–594.

Tolman, R. C. (1931). *Letter to Einstein*. The Albert Einstein Archives, The Hebrew University of Jerusalem. Archivnummer 23-031.

Tolman, R. C. (1934). *Relativity, Thermodynamics and Cosmology*. Oxford: The Clarendon Press.

Thorndike, L. (1949). *The Sphere of Sacrobosco and its Commentators*. Chicago: The University of Chicago Press.

Toomer, G. J. (1998). *Ptolemy's Almagest.* Translated and annotated by G. J. Toomer; foreword by Owen Gingerich. Princeton, NJ: Princeton University Press.

Tropp, E. A., Frenkel, V. Y. and Chernin, A. D. (1993). *Alexander A. Friedmann: The Man who Made the Universe Expand.* Cambridge: Cambridge University Press.

Van den Berg, S. (1996). Early history of the distance scale problem. In *The Extragalactic Distance Scale.* STScI Symposium Series No. 10, p. 1–5.

Van Maanen, A. (1916). Preliminary evidence of internal motion in the spiral nebula Messier 101. *ApJ*, **44**, 210–228.

Van Maanen, A. (1935). Internal motions in spiral nebulae. *ApJ*, **81**, 336–337.

Vibert Douglas, A. (1956). *The Life of Arthur Stanley Eddington.* London: Thomas Nelson and Sons Ltd.

von Zeipel, H. (1913). Recherches sur la constitution des amass globulaires. *Kungl. Svenska Vetenskapsakademiens handlingar*, **51**, No 5, 1–60.

Weyl, H. (1918). *Raum, Zeit, Materie: Vorlesungen über allgemeine Relativitätstheorie.* Berlin: Springer.

Weyl, H. (1919). Über die statischen kugelsymmetrischen Lösungen von Einsteins "kosmologischen" Gravitationsgleichungen. *Physik. Zeitschrift*, **20**, 31–34.

Weyl, H. (1921). *Raum, Zeit, Materie: Vorlesungen über allgemeine Relativitätstheorie.* Berlin: Springer, 4. Auflage.

Weyl, H. (1923a). *Raum, Zeit, Materie: Vorlesungen über allgemeine Relativitätstheorie.* Berlin: Springer, 5. Auflage.

Weyl, H. (1923b). Entgegnung auf die Bemerkungen von Herrn Lanczos über die de Sittersche Welt. *Physik. Zeitschrift*, **24**, 130–131.

Weyl, H. (1923c). Zur allgemeinen Relativitätstheorie. *Physik. Zeitschrift*, **24**, 230–232.

Weyl, H. (1930). Redshift and relativistic cosmology. *London, Edinburgh and Dublin Philosophical Magazine* (Ser. 7), **9**, 300–307.

Wirtz, C. (1922). Einiges zur Statistik der Radialbewegungen von Spiralnebeln und Kugelsternhaufen. *Astronomische Nachrichten*, **215**, 350–354.

Wirtz, C. (1924). De Sitters Kosmologie und die Radialbewegungen der Spiralnebel. *Astronomische Nachrichten*, **222**, 22–26.

Wollaston, W. H. (1802). A method of examining refractive and dispersive powers, by prismatic reflection. *Phil. Trans.*, **92**, 365–380.

Wright, Thomas (1750). *An original theory or new Hypothesis of the Universe, founded upon the laws of nature, and solving by mathematical principles the general phenomena of the visible creation; and particularly the Via Lactea.* Facsimile edition with an introduction by M. A. Hoskin. London: Macdonald, 1971.

von Zeipel, H. (1913). Recherches sur la constitution des amas globulaires. *Kungl. Svenska Vetenskapsakademiens handlingar*, **51**, No. 5, 1–51.

Zwicky, F. (1929). On the red shift of spectral lines through interstellar space. *PNAS*, **15**, 773–779.

Zwicky, F. (1933). Die Rotverschiebung von extragalaktischen Nebeln. *Helvetica Physica Acta*, **6**, 110–127.

Zwicky, F. (1935). Remarks on the redshifts from nebulae. *Physical Review*, **48**, 802–806.

Index